*Seafaring Scientist*

*Mayor at the helm of the* Physalia

# Seafaring Scientist

## Alfred Goldsborough Mayor, Pioneer in Marine Biology

*Lester D. Stephens and*
*Dale R. Calder*

The University of South Carolina Press

© 2006 University of South Carolina

Published by the University of South Carolina Press
Columbia, South Carolina 29208

www.sc.edu/uscpress

Manufactured in the United States of America

15  14  13  12  11  10  09  08  07  06    10  9  8  7  6  5  4  3  2  1

Library of Congress Cataloging-in-Publication Data

Stephens, Lester D.
   Seafaring scientist : Alfred Goldsborough Mayor, pioneer in marine biology / Lester D.
Stephens and Dale R. Calder.
       p. cm.
   Includes bibliographical references and index.
   ISBN-13: 978-1-57003-641-5 (cloth : alk. paper)
   ISBN-10: 1-57003-641-1 (cloth : alk. paper)
   ISBN-13: 978-1-57003-642-2 (pbk : alk. paper)
   ISBN-10: 1-57003-642-X (pbk : alk. paper)
   1. Mayor, Alfred Goldsborough, 1868–1922. 2. Marine biologists—United States—
Biography. 3. Zoologists—United States—Biography. I. Calder, Dale R. II. Title.
QH91.3.M39S74 2006
578.77092—dc22                                                    2006010192

Dedicated to Marie and Liz

# Contents

*Illustrations*

# Preface

Lying far out in the Gulf of Mexico, Loggerhead Key in the Dry Tortugas was once the site of a thriving marine biological station. The Tortugas Laboratory was founded a century ago by Alfred Goldsborough Mayor under the auspices of the Carnegie Institution of Washington, but only a few traces of the laboratory remain. Among the ruins stands a monument erected to the memory of Mayor in 1923, and this serves to inform the rare visitor to the islet of his scientific contributions. Yet, Mayor deserves wider recognition than is afforded by a bronze plaque located on a distant, windswept key, for he was an influential figure in American marine biology: the founder and director of the first tropical marine laboratory in the Western Hemisphere; a pioneer in the study of jellyfishes and coral reefs; and an early ecologist.

Scientists in Mayor's special fields remain keenly aware of his contributions, and several have named species in his honor. Others have provided useful sketches of the work of this able scientist and offered some interesting personal information about him, but we aim here to present a detailed account of his life and scientific contributions for both specialists and the general reader. We do not treat Mayor in isolation, however, but endeavor to place him in the context of his time and to view his work in light of contemporary scientific knowledge. Thus, our account intends to tell the story of a fascinating man and to show how he promoted scientific inquiries during the formative stages of marine biology. Alfred Goldsborough Mayor was a dedicated scientist, a talented artist, and a tireless laboratory director who effectively advanced the work of others.

*Acknowledgments*

Our work was greatly facilitated by a Senior Faculty Research Grant from the University of Georgia to Lester D. Stephens and grants to Dale R. Calder from the Natural Sciences and Engineering Research Council of Canada, the National Science Foundation's Partnerships for Enhancing Expertise in Taxonomy program, and the Royal Ontario Museum. We wish to express our gratitude to these institutions for their generous support.

Several archivists deserve our sincere thanks for their help. John Strom provided excellent assistance while we worked in the archives of the Carnegie Institution of Washington. Ellen Alers, assistant archivist at the Smithsonian Institution Archives, aided our research, as did Deborah Wythe, archivist and manager of Special Collections, The Brooklyn Museum; and Melanie Yolles, manuscript specialist, New York Public Library. The staff of the Department of Special Collections, the E. S. Bird Library, Syracuse University, readily filled our many requests for boxes of documents in their extensive collections of Mayor materials; and personnel in the Department of Rare Books and Special Collections, Princeton University Library, and in the American Philosophical Society Library were equally cooperative in making Mayor sources available to us. We are also grateful for the assistance of the staff of the University of Toronto's Gerstein Science Information Centre and Robarts Library Digital Studio, and the Royal Ontario Museum. Maureen Morin of the Digital Studio scanned a plate, prepared by Mayor, of the snail *Partula* from Tahiti, and Amanda Wagner assisted in reproducing the image of Mayor's plate of the ctenophore *Beroe*.

Niki Ryan and Willie Lopez of the National Park Service, Dry Tortugas National Park, made it possible for Dale R. Calder to visit Loggerhead Key twice, on March 6 and 7, 2004, and the organizing committee of the Carnegie Laboratory Centennial arranged for Lester D. Stephens to tour the Tortugas Laboratory site on October 13, 2005. Robert N. Ginsburg, Walter C. Jaap, and

Eugene A. Shinn were especially helpful in making the centennial celebration a memorable occasion, and we express our gratitude to them.

Dr. Anita Brinkmann-Voss, a longtime research associate of the Royal Ontario Museum, kindly translated for us a letter in German from Ernst Haeckel to Alfred G. Mayor. The four figures in this work, comprising a map and three diagrams of jellyfishes, were prepared by Patrice Stephens-Bourgeault of Studio Stephens-Bourgeault, Toronto. We deeply appreciate the assistance provided by all of the above-mentioned individuals.

A special word of appreciation is due to the late Brantz Mayor, who in July 2004, at the grand age of ninety-eight, clearly recalled for us memories of his father, Alfred Goldsborough Mayor. Equally helpful was his wife Ana Mayor, and together they welcomed us into their home and allowed us complete access to their large collection of Alfred Goldsborough and Harriet Hyatt Mayor papers. Contact with Brantz and Ana Mayor resulted from a chance conversation between Dale R. Calder and fellow biologist James T. Carlton, of Williams College and Mystic Seaport, who put us in touch with one of his former students, Naomi Gross, a great-granddaughter of Alfred G. Mayor. Naomi not only paved the way for us to meet her grandparents Brantz and Ana Mayor but also provided information about the Mayor papers they possess.

Readers and critics of a manuscript draft of this work have earned our lasting gratitude for their comments and suggestions, though they are in no way responsible for any errors or infelicitous expressions that may remain. We wish to thank Ronald L. Numbers of the University of Wisconsin for offering invaluable suggestions, especially on ways to broaden the context of our work. We are equally grateful to Jane Maienschein of Arizona State University for her encouraging support, insightful views, and thoughtful advice. Our thanks go also to Robert Buddemeier and Daphne Fautin of the University of Kansas, the former for reading the sections on coral reefs and the latter for a thoroughly detailed perusal of the entire manuscript. L. M. Passano of the University of Wisconsin provided constructive criticism of our comments on Mayor's contributions to jellyfish physiology.

Each of us is indebted to his spouse for her love and support, and we gratefully dedicate this book to them. Marie Ellis (Stephens) not only offered encouragement but also aided us by drawing on her skills as a longtime university librarian and by making insightful comments on ways to improve the work. Above all, she is a continuing source of inspiration. Elizabeth Mae Secord Calder battled cancer for more than six years before succumbing on January 23, 2004. To the end she encouraged and supported our efforts on this work. Her warm and considerate nature, courage, humanity, dignity, and love of family will not be forgotten.

*Seafaring Scientist*

# A Zeal for Zoology

Fascinated by a remarkable group of animals commonly called jellyfishes, Alfred Goldsborough Mayor spoke of "the rare grace of form and beauty of color of these creatures of the sea." In pursuit of knowledge about medusae, as biologists call them, he plied the oceans on many expeditions to collect, study, describe, and illustrate all he could find. In his voluminous work on the world's medusae published in 1910, Mayor summed up his labors: "Dry though these pages must be to the reader, to the writer they are replete with memories of the ocean in many moods, of the palm-edged lagoons of coral islands sparkling in the tropic sun, of the cold, gray waters of the northern sea bestrewn with floating ice, of days of withering calm in the heat of the torrid zone, and of adventure in the hurricane, all centering around the absorbing study of the medusae. Love, not logic, impels the naturalist to his work."[1]

As his career advanced, he continued to explore the seas and sea life, and he eventually expanded his interests to coral reefs and corals, which are related to jellyfishes. His ventures took him to remote oceanic islands in many parts of the world, and he praised their bounty and beauty in memorable words. A central figure in developing marine biology in the United States; an early ecologist; the founder and director of the Tortugas Laboratory, the first tropical marine station in America; and an authority on the South Sea Islands, he earned the esteem of his fellow scientists. Alfred Goldsborough Mayor was an adventurer, an artist, a writer, and a lover of nature, as well as a man of the seas. For him, no voyage in the interests of science was too risky or too long.

It had not always been so, for Mayor had not ventured out to sea until his twenty-fourth year of life. Born on April 16, 1868, at the home of his maternal grandfather, near Frederick, Maryland, he was named in part for his father, Alfred Marshall Mayer, and in part after his mother Katherine Duckett Goldsborough. For most of his life he spelled his surname "Mayer," but in 1918, during World War I, he altered it to "Mayor" in order to dissociate himself from his Germanic roots. Mayor never knew his mother, as she died sixteen days after giving birth to him. On June 30, 1869, his father married Maria Louisa Snowden. Before he did so, however, as his son later noted, he "went over the house destroying all traces" of Katherine Goldsborough Mayer. Alfred Mayor heard his father mention her name only once, and when he showed "any display of interest" in his mother, the inquiry was "met with stern repression." Fortunately for young Alfred, his stepmother filled the gap. Maria Snowden Mayer not only refused to allow her husband "to send [Alfred] off to some relatives" but also heaped great devotion on her stepson. She encouraged him to pursue his interest in natural history and praised his drawings. Maria bore two sons, Brantz and Joseph Henry, but the first died in 1874, halfway through his fourth year of life, and the second died in 1885, at the age of thirteen.[2] Alfred loved his stepmother, cared for her, and visited her often throughout his adult life.

His father was a self-taught physicist. Alfred Marshall Mayer had intended to pursue his interest in college, but his father, Charles Frederick Mayer, an attorney, wanted him to study law. Unwilling to accede to his father's request, however, Alfred Marshall quit school at age sixteen and went to work in a machinist's shop. Talented and a quick learner, he studied on his own, and in 1854, at the age of nineteen, he published a research paper in experimental physics. Other papers soon followed, bringing him to the attention of the University of Maryland, which appointed him as an instructor in 1857. He subsequently taught at Westminster College in Fulton, Missouri, from 1859 to 1861. Except for a brief period of study in Paris from September 1864 to April 1865, the activities of Alfred Marshall Mayer during the American Civil War are unknown. In a letter he penned on April 24, 1861, he made it clear that he thoroughly disliked Abraham Lincoln, unequivocally supported the notion of states' rights, and would take up arms if Maryland seceded from the Union. He did not enlist for service in the Confederate army, however. He resumed his teaching career in 1865 at Pennsylvania College (later Gettysburg College) in Gettysburg, Pennsylvania, and two years later moved to Lehigh University in Philadelphia.[3]

In 1871 Mayer accepted appointment as professor of physics at the Stevens Institute of Technology in Hoboken, New Jersey, where he remained until his death in 1897. Elected to the National Academy of Sciences (NAS) in 1872, Alfred Marshall Mayer enjoyed the esteem of his peers at home and abroad, counting among his friends Joseph Henry, John Tyndall, Baron Hermann von Helmholtz, and Louis Agassiz. In addition to the publication of ninety professional papers, Alfred Marshall Mayer published popular books on topics in physics and a book on fishing and hunting. Forgetting the situation he faced in the case of his own father and in the choice of a career, Alfred Marshall was determined that his son should follow in his footsteps.[4]

Acquiescing to his father's demand that he study engineering, sixteen-year-old Alfred Goldsborough Mayor enrolled in the Stevens Institute of Technology. Although he developed a "loathing" for engineering and physics, he excelled at mathematics and did especially well in chemistry. In 1889 Mayor received a degree in engineering, and despite his dislike of physics, he moved to Worcester, Massachusetts, in the fall of that year to pursue graduate work and serve as an assistant to Albert Michelson, professor of physics at Clark University. Unable to generate any enthusiasm for physics and longing to study biology, Mayor lost his position at the end of the academic year. Still under his father's sway, however, he began in 1890 to study advanced physics at the University of Kansas, serving as an instructor and assistant to Professor Lucian I. Blake. As a keen student of mathematics, he not only impressed Blake but also demonstrated an aptitude in physics by publishing the results of an experimental study on radiation and absorption of heat by leaves. The article appeared in the *American Journal of Science* in 1893. Four years later he published a description of a heliostat developed by his father. Since childhood, however, the first love of Alfred Goldsborough Mayor had been the study of natural history, and in March 1892 he decided to leave his position. Blake was sympathetic and wrote to Alfred Marshall Mayer to say that, while young Alfred was "successful in Physics, . . . his true taste and longings were toward natural history."[5]

The senior Mayer, "realizing at last that physics was innately distasteful" to his son, decided to support the effort of, in his words, his "strange and eccentric" offspring to enroll in the biology program at Harvard University. On February 19, 1892, then, Alfred Marshall penned a letter to Alpheus Hyatt, urging him to support the young Alfred's application for the study of zoology at Harvard University. A friend of the elder Mayer since their boyhood days, Hyatt was a noted naturalist, general curator of the Boston Society of Natural History, and a part-time curator in Harvard's Museum of Comparative Zoology (the MCZ). In addition he founded a seaside laboratory at Annisquam, Massachusetts, that eventually relocated to Woods Holl (now called Woods Hole) and became the highly respected Marine Biological Laboratory. Alfred M. Mayer also knew the director of the MCZ, Alexander Agassiz, and had delivered a series of lectures in 1874 at his summer school in marine biology, located then at Penikese Island, off the south coast of Cape Cod, Massachusetts. He may have suspected that Agassiz was away on one of his numerous collecting excursions or deeply involved in his mining enterprises, from which he had made a fortune. Thus, not wishing to delay the matter, Alfred Marshall Mayer wrote instead to Hyatt. Noting that his son "has a decided talent and great love for zoological studies," the elder Mayer said that his son desired to "give up Physics and devote his entire attention to zoology."[6]

No doubt, young Mayor needed no assistance from his father, for his record of academic excellence and his manifest interest in zoology were entirely sufficient to win admission for him. By then Mayor had already published an article in natural history. Appearing in the November 1890 issue of *Popular Science Monthly*, it was titled "Habits of the Box Tortoise." Although it offered no new scientific information, the account presented a good description of that reptile and included three excellent figures drawn by Mayor himself, one of which depicted two adult male turtles engaged in battle. As he had carved numbers in the plastron, or abdominal plate, of several individuals of the species, Mayor was able to provide data on the longevity of the turtles by recording the dates he had recaptured them. Meanwhile, he had drafted a popular article titled "Habits of the Garter Snake," which was published in the same magazine in February 1893 and which also contained drawings done by him. These simple accounts reflect the depth of Mayor's love of nature and his ability to observe the character and habits of its creatures. Later in life he recalled that, as a boy, "I threw myself heart and soul in the world of imagination . . . and sought my playmates among the creatures of the woods and field." As a child he "made colored drawings of nearly every butterfly and many of the moths" of his home region, and during his college years he devoted "almost every spare hour . . . to natural history . . . [and made] hundreds of colored drawings of turtles, snakes, newts, frogs, and insects." When he began to study biology, he was, he wrote, in a state of "ecstasy of delight and hope."[7]

Accepted by Harvard, Mayor moved to Cambridge, Massachusetts, in the spring of 1892. Although closely associated with Alexander Agassiz as an assistant in the MCZ, he received considerable direction and support from the young zoologist Charles Benedict Davenport, who had only recently received a doctorate from Harvard and was serving there as an instructor of physiology. Davenport also taught courses in "experimental morphology" and "causes of variation." Mayor chose to work with the zoology professor Edward L. Mark, a noted morphologist, as his dissertation director, and for his topic he elected to study the development and coloration of wing scales in butterflies and moths. The "real instigator" of the dissertation topic, however, was Davenport. In addition to his ability to do statistical studies of Lepidoptera, Mayor proved to be adept at drawing the wing scales. The subject and nature of his studies were likely of no great interest to Agassiz, who concentrated instead on marine invertebrates. An old-fashioned naturalist who generally frowned on modern biological approaches, Agassiz nevertheless recognized Mayor's skill as an artist, his ability to learn quickly, and his zeal for zoology.[8]

Mayor had actually begun his work by first studying at Agassiz's marine laboratory in Newport, Rhode Island, during the summer of 1892. His ready mind and his ability to illustrate jellyfishes accurately so impressed Agassiz that the noted zoologist soon invited him to coauthor a book with him on the medusae of the eastern coast of North America. Honored by this request and eager to contribute to science, Mayor accepted the invitation, and early in January 1893 he was off on an excursion with Agassiz to waters around the Bahama Islands to collect, describe, and draw medusae inhabiting the western Atlantic Ocean. On January 8, aboard the steamer *Wild Duck,* Mayor began a venture that would consume his interests for years to come.[9]

Most of the sampling of the expedition was carried out in shallow water until February 3, when the crew dropped the net to a depth of 300 fathoms (549 meters). The haul also included animals captured near or on the surface. Among the latter was a beautiful specimen of the Portuguese man-of-war, *Physalia physalis.* Mayor set about immediately to draw that siphonophore while it was still alive, taking special care to avoid contact with the long tentacles that contained batteries of powerful nematocysts, or stinging capsules. Mayor had put the finishing touches on the drawing, or on another one of *P. physalis,* by March 12, 1893, for the original, dated version remains extant.[10] The drawing, colored in true representation, exhibits the skill of Alfred Mayor with pen and brush.

Agassiz allowed Mayor to draft a report of four species of hydromedusae and two of siphonophores collected on this trip. As was the case with his father, Louis Agassiz, Alexander Agassiz strictly regulated publishing by students and assistants connected with the MCZ, being certain that his name was associated with any publication resulting from specimens obtained for the museum or, at the least, that the subordinate had obtained his explicit permission to publish a paper. In this instance he permitted Mayor to proceed as sole author of "An Account of Some Medusae Obtained in the Bahamas." Perhaps he did so because of the press of other business, but for whatever reason, it is clear that Agassiz knew that Mayor was fully up to the task. Certainly, he must have admired Mayor's drawings of the six species described in the paper. Published in the *Bulletin of the Museum of Comparative Zoölogy* in 1894, the paper assured that Mayor was on his way to becoming an authority on the medusae. Of the six species addressed in his report, he described four hydromedusae as new to science. Three of the names he established remain valid: *Bougainvillia niobe* and *Hybocodon forbesii* (= *Vannuccia forbesii*), of the order Anthoathecata; and *Cubaia aphrodite,* of the order Limnomedusae. Both orders belong to the class Hydrozoa (phylum Cnidaria). In general, hydromedusae are smaller than scyphomedusae, which belong to the class Scyphozoa, or "true jellyfishes." The hydromedusae, or "veiled medusae," are fundamentally different in that each normally possesses a velum, or a small shelflike veil on the inner margin of the bell or umbrella, and they lack cells in their jelly. Mayor's first article on medusae solidified Agassiz's determination to work with Mayor in producing a comprehensive, systematic treatment of medusae of the Atlantic coast of North America, though Mayor later expanded the scope to encompass the medusae of the world.[11]

Meanwhile, by late fall of 1893 Mayor was experiencing strain in his left eye, and on a physician's advice, he spent the next six months in a darkened room at his parents' home. His stepmother kept up his spirits by reading to him from "an adjoining room." Mayor recovered from the setback, however, and, after a trip to Europe with his father and stepmother during the summer of 1894, returned to his studies. He pushed forward with his academic courses and with his investigation of the nature and color of wing scales in Lepidoptera. The work on butterflies and moths proved to be very time-consuming, as it was based on careful microscopic investigations, and, quite pleasing to Davenport, employed statistical analyses, which included calculations by hand of seventy-seven hundred variations. Mayor made excellent, detailed drawings of various types of wing scales.

Field trips also kept him busy. On one of these, at a location he did not specify, he collected several specimens of hydroids, which are the sessile or polyp stages in the alternating life cycle of the medusae known as anthoathecates and leptothecates (see chapter 4, p. 84, for more detail on the remarkable life cycles of hydromedusae).[12] Although the notion of alternation of generations was understood by that time, zoologists had much to learn about which hydromedusa belonged to which hydroid. As a consequence, a hydroid, or polyp, was often given one name while that of its medusa received another. With his focus on medusae, Mayor paid relatively little attention to hydroids, but he thought he had discovered three new species of them in 1895. None of his hydroid taxa proved to be valid, however.

In addition to his scientific work, Mayor was beginning to spend more and more time in the home of Alpheus Hyatt as he was attracted to Hyatt's daughter Harriet Randolph, who, like her younger sister Anna, possessed exceptional talent in art, especially sculpting. Indeed, Harriet had already won recognition for her works of sculpture when she first met Mayor. By 1896 Mayor had declared his love to Harriet and asked her to marry him. Harriet had long admired the aspiring zoologist, who stood only 5 feet and 7 inches (1.7 meters) in height but made a striking figure with his slender physique, attractive blue eyes, and charming manners. She readily returned his love.[13] The couple chose, however, to keep their engagement secret and to delay marriage until Alfred had completed his work at Harvard and had a job that insured sufficient income. Moreover, both feared a breach with Agassiz if Mayor should seem to be paying less than full attention to his duties. Agassiz had plans for Mayor for the next few years, including the production of a major portion of the projected book on medusae. Alfred and Harriet believed that it would be unwise to cross the aging autocrat at this point.

Mayor hoped to complete his monograph on butterflies and moths before Agassiz's next scientific expedition in February 1896 and to publish it in two parts before the date of departure. Unfortunately, as neither was quite ready for publication, he had to entrust the task of revision to the capable hands of his dissertation director, Edward L. Mark, who edited many articles for the museum's *Bulletin.* Mayor's articles appeared in volumes 29 (1896) and 30 (1896–97) of the *Bulletin,* respectively as "The Development of the Wing Scales and Their Pigment in Butterflies and Moths" and "On the Color and Color-Patterns of Moths and Butterflies." These studies drew on Mayor's broad knowledge of physics, chemistry, biology, and statistical procedures, and they clearly reflected his acceptance of the role of natural selection in the development and coloration of lepidopteran wings. Accompanying the first part are seven plates of seventy-four figures of wings and wing scales of Lepidoptera, and with the second part are four plates of sixty figures of moth and butterfly wings—all neatly drawn by the hand of Alfred Mayor.[14]

Although his fascination with Lepidoptera would continue, Mayor turned his attention once again to medusae, a subject that would consume much of his professional time for the next few years. On February 27, 1896, along with Agassiz and his son Maximilian and Dr. William Woodworth, a young member of the MCZ staff, Mayor boarded a train bound for the West Coast. The group arrived in San Francisco on March 4 and three days later set sail for Australia to collect specimens of invertebrates, including, of course, jellyfishes inhabiting the waters of the area. The ship stopped for a short time in Honolulu and again briefly in Apia, Western Samoa. Mayor was immediately fascinated with the native inhabitants of both places. In the journal he kept on this trip, he described the form of the Polynesians as "superbly symmetrical," and he praised their "beautifully poised heads" and their "clear rich brown" skin. In their complexion, he added, there was no sign of "sooty color of the negro or papuan [*sic*]." To Mayor, "all of their [the Polynesians'] muscles were evenly developed." These initial observations on Polynesians as the ideal body type and, indirectly, on the supposed inferiority of peoples of darker skins, reflected his belief in the evolution of humans, and in due course Mayor would write at length about this idea. Few of his fellow travelers went as far as he did, however, in praising

Polynesian beauty. Mayor was struck, moreover, by what he considered to be the virtual enslavement of those peoples by the imperial powers, and from the outset he lamented the destruction of their culture and the demise of their "happiness and purity."[15]

The expedition party put ashore briefly at Auckland, New Zealand, where Mayor got a glimpse of another group of Polynesians, the Maoris. On April 2 the ship entered port at Sydney, Australia, and for ten days Mayor explored the surrounding country, recording in his journal detailed descriptions of the terrain and of some of the reptiles and metatherians, or marsupial mammals, of the region. On to Brisbane the group sailed, and on April 16, aboard the vessel *Croydon,* the scientific part of the expedition commenced in earnest. It lasted for almost two months. In his journal Mayor described surrounding coral reefs and commented on their formation. Eventually he would explore coral reefs more thoroughly and form some useful hypotheses about their formation and ecology. Meanwhile, the nets of the *Croydon* were bringing up considerable numbers of invertebrate specimens, but because of strong winds, the hauls included relatively few medusae. In fact, as later reported by Agassiz and Mayor in "On Some Medusae from Australia," only one of the two scyphozoans they collected was thought to be new, but even it turned out not to be so.[16] In any case, although the name of Agassiz appeared as senior author of the article, Mayor almost certainly wrote the descriptions of both species. He drew superb figures of each, which added to the portfolio of illustrations that would later be used in the book on the medusae of the world.

Early in June 1896 the expedition left Australia, bound for Naples, Italy, via Ceylon and the Suez Canal. On July 3 the group arrived in Naples, where Mayor likely visited the renowned Stazione Zoologica di Napoli, which was then a leading institution in marine zoology. Embarking for England soon thereafter, thence to the United States, the expedition ended 105 days after it had begun. By early October, Agassiz had officially given Mayor the duties of curating the Radiata (the name of an assemblage, now known to be artificial, that included Cnidaria, Ctenophora, and Echinodermata) and otherwise assisting him in the work of the MCZ, for which Mayor was to be paid one thousand dollars annually. While it was low-paying, the position offered Mayor an opportunity to strengthen his reputation as an authority on the medusae and kept him close to his beloved Harriet. His impressive record of research and publishing now evident and his dissertation completed, Mayor received a letter from Davenport on October 30, 1896, notifying him that he had been recommended "to the Faculty of Arts and Sciences for the degree of Doctor of Science in Natural History; Subject Zoology."[17] Soon thereafter Harvard bestowed the degree on Mayor.

In the meantime, Alfred Mayor had begun to draft another article on Lepidoptera. It was published in two parts in *Psyche,* a leading journal in the field of entomology, in April and May 1897, respectively.

Titled "A New Hypothesis of Seasonal Dimorphism in Lepidoptera," the first part of the article provided a thorough review of the literature on the subject, while the second part posited the hypothesis that the process of "natural selection . . . operate[s] to cause all summer pupae [of Lepidoptera in temperate zones] to *inherit* a low metabolism." In Mayor's view, the metabolic rate of Lepidoptera inhabiting warm regions should be high in order to assure rapid development, but that of the pupae produced by such species during the fall season should be low so that they could survive warm autumns without emerging. Clearly, Mayor accepted the theory of evolution by natural selection, though, as did others of his contemporaries, he subscribed in part to the neo-Lamarckian view of evolution as a consequence of species adaptations to environmental conditions. Inasmuch as he never penned his full views on the subject, however, it is difficult to determine how he reconciled seemingly competing arguments. As the historian of science Ronald L. Numbers has clearly shown, a welter of positions on the causes and nature of evolution existed at the time, and it is essentially hopeless to try to categorize contemporary evolutionists. In any case, Mayor was certainly an evolutionist, and he equated Charles Darwin's work with that of Isaac Newton.[18]

The topic of evolution was not one he felt free to discuss with Agassiz, however, for his mentor remained ambivalent about it, perhaps in part out of respect for his father, Louis, who had strongly opposed it. As Mary P. Winsor notes, Alexander Agassiz was never enthusiastic about the theory, and he disliked the "zealous tone of committed evolutionists." Moreover, Agassiz viewed the goal of research in natural history as primarily taxonomic, and he generally shunned theorizing of any kind in science. Mayor, on the other hand, subscribed to what he and many others called "modern biology."[19] To them, descriptive taxonomy was a part of systematic biology but not the sole aim of the science.

In the spring of 1897 Agassiz sent Mayor to collect medusae around the remote keys of Florida, where Agassiz's former assistant Jesse Walter Fewkes had successfully taken many specimens several years earlier. It is likely that the young zoologist was eager to see what he could collect there, and as it turned out, the experience was enormously rewarding for him. On June 1, 1897, Mayor wrote to Harriet from Key West, a small city that he found filled with unsavory characters, or "sharks," as he called them. "The squalor of this place is disgusting," he said, and he drew a charming sketch of himself in a net-covered bed surrounded by huge mosquitoes and a large rat. Nevertheless, he declared that he had collected three new species of medusae, although he hoped for even better luck in the westernmost keys known as the Dry Tortugas, or simply Tortugas. In fact, none of the three species was new, but Mayor did indeed have better luck at Tortugas, where he collected several new species of hydromedusae, including the anthoathecate *Bougainvilla frondosa,* and the leptothecates *Oceania gelatinosa* (=*Clytia gelatinosa*) and *Oceania globosa* (=*Clytia globosa*).[20]

Soon Mayor arrived on Loggerhead Key, the largest island of the Tortugas group but at that a mere 3,950 feet long and 600 feet wide (1204 x 183 meters). On this small, sandy key stood a United States lighthouse, whose keeper and family befriended the lonely researcher. For more than two weeks Mayor arose at 4:30 A.M. each day and rowed out "in a small dingy" to collect medusae. In a letter to Harriet written on June 20, he boasted that he was collecting as many as three or four undescribed species of medusae each day. That claim proved to be an exaggeration, but, in fact, Mayor was meeting with great success. Moreover, he became fascinated with "this little speck of an island far out in the ocean." Rowing above the surrounding coral reefs, Mayor told Harriet, he could see "the purple and yellow sea fans slowly waving in the surges . . . the ocean's graceful lacework." He added, "the breezes here are simply heavenly." It was, he imagined, an ideal place for research in marine zoology—perhaps even for a permanent marine biological station.[21]

Not long after he returned to Cambridge, Mayor received word that his father was near the point of death, and he rushed to Maplewood, New Jersey, to be at his side. Alfred Marshall Mayer died on July 13, 1897, which, according to his son, came "after an illness . . . [of] some months." In a letter to Harriet a few days later, Mayor praised his late father, noting especially his ability to solve "abstruse problems in Physics."[22] Less than a week later Mayor was preparing for two more collecting trips—the first to the coast of Maine and to Grand Manan, an island located in Canadian waters at the entrance to the Bay of Fundy, and the second to Fiji. By mid-August he was in Eastport, Maine, at the northeastern corner of the state, collecting various marine worms, brachiopods, and, of course, medusae. He experienced less success than he had hoped for, but by the end of the month he had brought in several specimens, including, he claimed, a species of medusa not seen in more than three decades, though he did not identify it. Soon thereafter Mayor decided to cancel the trip to Grand Manan, New Brunswick, and go to Charleston, South Carolina, instead.[23]

Meanwhile, he told Harriet that "Alexander the Great," his private name for Agassiz, had set October 8 as the date of departure for Fiji. A week later he informed her that Agassiz had offered to withdraw as coauthor of a nearly completed article on the scyphomedusa known then as *Dactylometra quinquecirrha.* Later referred, or moved, to the genus *Chrysaora,* this is a venomous species widely known along the United States East Coast as "the sea nettle." Agassiz told

Mayor that since he had done "all the work," Mayor should therefore be listed as the sole author. Mayor apprised Harriet, however, that he had declined to remove the name of the "truly generous and magnanimous" man. More than the kind offer by Agassiz motivated Mayor to insist on listing his boss as co-author, however, for as Mayor told Harriet, "as long as he furnishes the funds and the encouragement, he shall share the 'glory.'"[24]

The reason for Mayor's decision to forgo Grand Manan Island for the harbor of Charleston is unknown, but it is probable that he expected to be more successful in collecting there, especially species of hydromedusae and ctenophores, or comb jellies. Taxonomists in Mayor's day lumped the ctenophores with the cnidarians into the phylum Coelenterata, though the comb jellies differ in some attributes, including the lack of nematocysts, or stinging capsules. In any case, Mayor was certainly aware of the work of the late John McCrady of South Carolina, one of North America's pioneering hydrozoan zoologists, a student of Louis Agassiz during the 1850s, and a colleague of Alexander Agassiz at Harvard from 1873 to1877. A native of Charleston, McCrady had collected and described many species of hydromedusae from Charleston harbor during the mid-1850s. He had also collected numerous specimens of ctenophores there and had established two new species of these beautiful creatures, *Bolina littoralis* and *Beroe punctata*. Mayor referred to McCrady's noted paper on hydromedusae as "a classic in Zoological literature." He even sought out Edward McCrady, the oldest brother of John McCrady, in order to pay tribute to the late zoologist.[25]

Mayor found the city of Charleston charming, but he described it to Harriet as "the city of faded grandeur, abounding . . . in proud old houses of the vanished days of splendor when cotton was king." He admired the "great houses" in the city, and he noted the statues to Confederate "fallen heroes who defended Charleston" during the American Civil War. Among the city's institutions, one impressed him especially. It was the Charleston Museum, one of the oldest of its kind in the United States. The museum, he observed, "is surprisingly good, the collection being quite extensive and very well kept." The building, however, was poorly lighted, he added. Most of all, Mayor was pleased over the abundance of medusae in Charleston harbor. "I am head over heels in jellyfishes," he told Harriet. "Such a hawl [*sic*] never was thought of. It beats Tortugas."[26]

After a week of successful collecting in Charleston during the late summer of 1897, Mayor headed for New Jersey and Maplewood to visit his stepmother. By then Harriet was growing a bit impatient over the long absences of her "dear old man," as she often affectionately referred to Mayor, who, like her, had now reached the age of thirty. Even more, she had come to believe that Agassiz would use Mayor as long as he could keep him and never offer him a permanent professional appointment. Her fears were not unfounded, for Agassiz habitually kept his assistants in a subservient position. In fact, Harriet wrote to Mayor, "The lesser Agassiz, your 'Alex the Great,' is to me Ivan the Terrible," and she referred to Agassiz as "spider-like" and ready "to absorb all the men who fall in his great web, the Museum." The percipient woman was indeed close to the mark. Unlike her fiancé, she suspected a scheme behind Agassiz's offer to allow Mayor to be the sole author of the paper on *Dactylometra*. Mayor told her not to worry, however. He added, "I grant freely that he is a grasping, avaricious business man," but he believed that Agassiz had been "perfectly frank and generous" toward him. It "would be most unwise," he informed Harriet, for him to withdraw from the project on the medusae. "The best thing to do," he suggested, "is to finish it and then try for a position at Harvard if possible, but, if not, in some civilised [*sic*] eastern city." He was not afraid, he said, that Agassiz would "attempt to 'gobble up' my work, for he could not do it." Besides, he noted, "250 of the drawings are my own personal property," and he observed that, as Agassiz had never seen "many of the species, . . . he could give but imperfect descriptions of them if he tried it alone." While Agassiz kept tight control of his assistants and MCZ publications, he did not intend to deprive Mayor of the projected work.[27]

Mayor and Harriet enjoyed but a little time together, as Mayor had to leave with Agassiz and William Woodworth for the West Coast on October 8, 1897, via Vancouver, British Columbia, where they arrived on the 14th. Four days later the party was aboard a ship headed for Honolulu. Mayor had developed a thorough dislike of Woodworth during the Australia expedition, viewing him as a sycophant who hoped that Agassiz would make him his successor as director of the MCZ. Mayor ignored Woodworth as much as he could, but his regard for the man had changed little by the time of this trip. "Little Billy remains in sight," Mayor told Harriet, "but is extremely sulky." He viewed him, however, as "much more manly and not so utterly clownish and insignificant as he was on the Australian trip." At no other time in his life did Mayor express such disdain for a person. In part, the tension was probably due to close proximity on a long voyage. It was also likely a consequence of Mayor's realization that Woodworth enjoyed greater status in the hierarchy of the MCZ, or, more aptly, in Agassiz's scheme of things. Moreover, Mayor was now concerned about the capacity of the aging Agassiz to continue to work effectively. In a letter to Harriet written aboard ship, he said, "Mr. Agassiz is beginning to show his age . . . [and] lacks much of his former energy."[28]

Mingled with these musings were happy thoughts of the marriage planned for the summer of 1898. Mayor declared to Harriet that he would "tell Alex the huffy" of their plan in April. He had been thinking about the nature of their wedding ceremony and said, "I have made up my mind that you shall not be obliged to promise 'to serve and obey' . . . [for] it is repulsive to me that you should be obliged to utter menial words that neither of us intend to respect." To forgo the traditional phrases, however, Mayor thought it might be necessary to get a Unitarian minister to perform the ceremony.[29] Neither Alfred nor Harriet held any strong attachment to traditional religion, and both believed in equality of partners in marriage. Their hope for a wedding in 1898 was overly optimistic, however.

In Honolulu on October 28, 1897, Mayor began a long letter to Harriet that he did not finish until a day after he arrived in Suva, Fiji, on November 8. Once again he expressed great admiration for "the beauty of the natives" of Hawaii, whom he nevertheless viewed as "mere withering remnants of a once proud race." Mayor extolled "their rich bronze skins and smiling yet pensive faces framed by luxuriant locks of rippling black hair." Despite his admiration of the native Hawaiians, however, Mayor believed that they possessed "the minds of children." Yet, he deplored their mistreatment by Americans and the introduction of "hideous diseases" by "our 'superior' race." Among the indignities heaped upon the native Hawaiians, Mayor observed, was the buying up of all the beaches by Americans, who then charged the traditionally devoted swimmers a fee to go into the water. "It is," he asserted, "but a last act in the disgraceful drama that our government should 'annex' them in order that a few sugar planters and politicians may gain more gold." Still, Mayor found the setting beautiful, saying to Harriet, "all around us lies an ocean of brilliant blue flecked with purest white where the trade wind ruffles it." He added, "at night in the moonlight it gleams like molten silver under the deep soft purple of the sky."[30]

Far less pleasing to Mayor was the Fijian capital. He described Suva as "a small, struggling, uninteresting town," and he said that "the streets swarmed with natives." Yet, he depicted the native Fijians as "great superb looking men and women. . . [whose] complexions are dark and muddy in color . . . [and whose] hair stands straight out like a mop." On the whole, Mayor was not displeased with Fiji, especially as he was successful in collecting, as he reported to Harriet, "many new jelly fishes." In mid-December, Agassiz extended the expedition of the steamer *Yaralla* to the waters off Sydney, Australia, for a week, and Mayor happily told Harriet that "the trip has been from a scientific standpoint the most successful that Mr. Agassiz has as yet undertaken." He noted, however, that the officers of the ship "hate Mr. Agassiz . . . on account of his having quite naturally made a hog of himself" in collecting specimens. Woodworth, he added, "is thoroughly dispised [*sic*] and held in universal contempt," a consequence of his constantly "fawning over Mr. Agassiz." Others shared Mayor's view of Woodworth, and even Agassiz sometimes "found his agreeable nature exasperating." Mayor expressed particular dislike of Woodworth because, as he viewed the matter, Woodworth exhibited special "insolence" toward him. On the positive side, Mayor enthusiastically reported that he had collected "about 30 new jellyfishes, which will make a good paper."[31]

The expedition was back in Fiji on December 24, and it remained there into January 1898. By then Mayor was missing Harriet greatly, but he found several letters from her awaiting his arrival. The letters, he wrote to his beloved, "gave me a great big lump in my throat . . . and I felt like blubbering in a whaly fashion [as] I was so happy." He urged Harriet, who, even more than he, often misspelled words, to be careful when writing the name of Agassiz. She had sent her letters to Alfred in care of "Alex Agazzie." If Agassiz had seen the envelopes, said Mayor, his (Mayor's) career might have been "wrecked . . . for [Agassiz] is a deep old fox." Meanwhile, as he bided his time before the scheduled date of departure on January 13, Mayor was in a mild state of depression. On New Year's Day he wrote to Harriet that "heavy black clouds brood over the gloomy mountains, reeking with the moisture of many a day of rain." Worse, in his view, were "the swarms of bushy headed black Fijians, and [the] mean, sneaky looking, dark brown Hindoos [who] roam lazily through the ugly streets of stupid Suva."[32] Had the natives been the lovely, bronze-skin Polynesians whom he so admired, Mayor might have been more cheerful. In any case, his bias toward peoples of dark skin was clearly a part of the idea of racial gradations that he was forming —and it differed little from that of many of his western Caucasian contemporaries, whether they accepted the differences as part of a divine design or, like him, as a function of evolution.

A brief stop in Honolulu on the return voyage gave Mayor an opportunity to pen a letter to Harriet, and in it he said, "A. Ag[assiz] thinks he has me by the wool" and that he must find "a more lucerative [sic]" position as soon as possible. Back in San Francisco by mid-February, Mayor departed for home on the twenty-first day of that month. Before he left the West Coast, however, he wrote a more optimistic letter to Harriet, saying that he and "old Moloch [Agassiz] . . . always agree exactly" and adding that he had "just persuaded him to found a biological laboratory at the Tortugas."[33] After he got home and talked over matters with Harriet, however, Mayor began to fret over his modest income and the improbability that Agassiz would give him a raise. It was already clear to the couple that they could not afford to get married during the coming summer as they had planned. Moreover, Mayor was worried about Harriet's constantly sore throat, a symptom often associated with tuberculosis, though neither he nor Harriet was willing to admit that possibility. Prompted by these concerns, Mayor composed a letter to Agassiz on February 28, saying that he would be unable to continue to work for him after next October, for his income was inadequate. The situation would therefore make it impossible for him to work on the medusae book, but he would, during the coming summer, "devote all of my energies toward advancing the work . . . and will then leave the material with you." Mayor expressed regret that he was "obliged to abandon this work," and he thanked Agassiz for his "constant generosity and kindness."[34]

Unable to bring himself to make a break with Agassiz, however, he never sent the letter. A low income was better than no income, although in reality, as he later noted, he could have become an engineer. He detested that profession as much as he loved zoology, however. Moreover, he could not bring himself to abandon the book project. Not only had he invested an enormous amount of time and energy in it, but he also knew that Agassiz lacked the stamina, time, and detailed knowledge of recently discovered medusae to bring the book to completion. In addition Mayor continued to hope that his success in publishing would eventually win for him a regular appointment at Harvard. In fact, the study of *Dactylometra* was scheduled to appear in the MCZ's *Bulletin* within a few weeks.

As previously noted, the article titled "On *Dactylometra*" carried the name of Agassiz as first coauthor, but it was written by Mayor. In addition Mayor drew thirty-four detailed illustrations to accompany the article, and their accuracy and artistic quality left no doubt that he could capture the fine characters of his subject. The specimen from which the drawings were made had been collected in Cuba's Havana harbor on February 23, 1893. Mayor included a photograph and a drawing of *Dactylometra* done by others, but he did all of the remaining figures.[35] The authors, or really Mayor, argued that *Dactylometra* differed from *Chrysaora,* but as later authors established, the animal is now known as *Chrysaora lactea.* The perceived differences between the two supposed genera are due entirely to variations representing different growth stages. Mayor based his judgment on the number of tentacles possessed by the specimens collected, which happened to be forty, rather than the twenty-four associated with the type species of the genus *Chrysaora.* In fact, the number of tentacles on, as well as the coloration of, *C. lactea* is quite variable. In any case, Mayor provided a thorough account of the morphology and habits of the species, further indicating that he was now a competent authority on the medusae.

Either Agassiz or Mayor, but likely both, decided that more collecting should be done at Tortugas, and by the first of July 1898 Mayor was back on Loggerhead Key. Again taken by the splendor of his surroundings, he told Harriet, "We have the most glorious sea and sky in the world here, not excepting Naples." Quite clearly, Mayor was envisioning the establishment of a marine biological station on the small island—one that, he hoped, would rival the great Stazione Zoologica in Naples, Italy. Enthralled by the setting, Mayor wrote to Harriet, "The ocean glistens in splendid emerald and deep blue, and great cumulus

clouds drift over . . . in the afternoon, towering far upward into the blue." Each
day, he added, "I sail for hours . . . over the rippling waters and look far down
into the recesses of the coral caverns where the cool deep shadows invite one to
plunge beneath our prosaic world into the brilliant enchanted one below." His
collection of "new medusae" was growing as well, and he was, he informed Harriet, making many photographs of the Tortugas. A letter from Harriet, dated
July 10, dampened Mayor's enthusiasm, however. "Agassiz," she reported,
"has appointed Woodworth his successor—or rather his dummy, for that is all
he can be with 'A. Ag.' pulling the strings." Mayor expressed concern once again
that he had no prospect of earning more than one thousand dollars a year if he
turned elsewhere, unless he went into the field of engineering, which, he added,
"I hate."[36]

In August 1898 Agassiz officially retired as director of the MCZ, but he
retained de facto control of the institution. Woodworth became "Assistant in
Charge" of the museum, an appointment that convinced Mayor that he now
had no hope of a permanent position there. His dream of someday becoming
director of the MCZ was dashed. "Mr. Agassiz," he wrote to Harriet, "never
would have appointed me to such a position; indeed he cares nothing for me
personally and our relation is only a scientific alliance." Mayor might have accepted the appointment of someone else as Agassiz's successor, but he could not
abide Woodworth, of whom he had "formed a very unfavorable opinion . . . as
a man of honor and a man of science." Mayor added that Agassiz was "very fond
of Woodworth, who has invariably made himself as polite and serviceable as
possible," although Woodworth was prone to "abusing poor Alex behind his
back and saying . . . the Museum will be better off when he [Agassiz] is dead."
For the time being, however, Mayor would have to go on, though he said he
hoped he would be able to resign before the coming winter. Meanwhile, Mayor
continued to collect medusae around Tortugas, and by early August he could inform Harriet that he now had "a very good idea of the character of the Medusan
fauna of this region." He was also excited over the discovery of a polychaete
commonly called the palolo, and he made notes that would, at some point,
allow him to publish a study of this unusual worm.[37]

Mayor left Loggerhead Key on August 16, but an outbreak of yellow fever in
Key West forced him into quarantine in the city he viewed as filthy and stink-
ing. When he finally got to Charleston in late August, he stayed briefly, perhaps
to collect more medusae. This delay disappointed Harriet, especially since her
fiancé was due to meet Agassiz in Eastport, Maine, in mid-September for additional collecting in northern waters. No doubt Mayor wanted to be with Harriet; after all, he had been gone a long time, and they had been compelled to
postpone their wedding. Nevertheless, Alfred was absolutely driven to search for

yet more new species of medusae. He became so "homesick" in Eastport, however, that he actually decided to go to church. Reared in the Episcopal faith but never confirmed, Mayor thought first of attending a Unitarian service but concluded that the group looked "too intellectual." He settled for an appearance in the "shouting Methodist" church. This was the first time he had "gone voluntarily to church in ten years." The experience evoked memories of the years "when my mother [stepmother] began to teach me about Hell," he said. Had she "dwelt on heaven instead," he added, she might have succeeded in keeping him interested in Christianity. In any case, Mayor viewed the sermon as "a lark," but the religious service at least assuaged his feeling of loneliness. For Alfred Mayor, religion remained a childhood remembrance of God as "a terrible, heartless, vengeful, and exacting tyrant," but except for later criticisms of Christian missionaries in the South Sea Islands, he appears to have avoided the subject of religion in personal communications.[38]

In the meantime, Mayor had received additional word from Harriet about the situation at the MCZ, along with Harriet's views on the matter. "Any definite connection [with Agassiz] . . . after the last few developments," she said, "would be disagreeable to me, and there will be more to come with Woodward [Woodworth] as [a] tool." She noted that Agassiz had offered Woodworth's previous position to Walter Faxon, an able naturalist who had been associated with the museum for more than two decades. Faxon declined to accept the offer, and according to Harriet, the vengeful Agassiz immediately reduced the unfortunate man's salary.[39] This news served to confirm the view of Alfred and Harriet that Mayor must continue to be cautious.

Mayor knew that he could hardly afford to decline to accompany Agassiz on another trip to the South Pacific, planned for late 1899. "I will simply have to go, and I feel wretchedly about it," he wrote to Harriet. At least Woodworth would not go on that expedition, he cheerfully added. Resigned to the situation, although he toyed briefly with the idea of seeking a position at the Imperial University of China, Mayor devoted the next few weeks to completing a study of the medusae collected around Fiji in late 1897 and 1898. With Agassiz's name again listed first and Mayor's second, the article, titled "Acalephs from the Fiji Islands," appeared in the February 1899 issue of the *Bulletin of the Museum of Comparative Zoölogy.* Once again Mayor wrote the descriptions, and he drew and colored fifty-four figures, contained in seventeen plates. Of the thirty-eight species they described, Agassiz and Mayor believed that twenty-four were new. Eventually, however, some were determined to be synonyms, or names given to species previously described by others. Yet, thirteen of the Agassiz-Mayor nominal taxa of hydromedusae, two of rhizostome scyphozoans, and two of ctenophores are considered valid today. The siphonophore they described as new turned out not to be so. Mayor chose native words to name some of the species, describing the bluish-colored rhizostome jellyfish as *Cephea dumokuroa* (=*Netrostoma dumokuroa*) after "the native Fijian name for this species," which he found in "a swarm" on November 25, 1899, but encountered no more of thereafter on the voyage. One of the ctenophores he discovered, *Eucharis grandiformis* (=*Leucothea grandiformis*), appeared "in considerable numbers," but the delicate specimens had to be handled carefully as "the least touch . . . [was] sufficient to tear the tissues of the animal."[40] The monograph on acalephs from Fiji represented a major contribution to the field of coelenterate biology.

Ironically, this important monograph, along with other systematic studies by Mayor, was perhaps less appreciated by many contemporary biologists than it should have been, as taxonomic work no longer held a prominent place in many academic institutions. Old-style natural history was giving way to so-called modern biology. Following the model established at Johns Hopkins University in the mid-1870s and the development of disciplinary societies in the field of biology, interest in traditional natural history was fading, and along with it, taxonomic studies. Alexander Agassiz did not follow the trend, however, and so his student Alfred Mayor followed suit, though he was already beginning to show signs of breaking from the mold. Nevertheless, identifying, naming, and classifying species remained basic to developments in biology, and they were especially important in the study of cnidarians, about which so much was yet to be learned.[41]

From late in 1898 into May 1899 Mayor worked hard to keep the book project on schedule. Particularly concerned over the systematics of the hydromedusae, he labored alone, for Agassiz apparently devoted little, if any, time to the effort. In this, Mayor was fortunate because he was much more up-to-date in his understanding of the group than his sponsor was. On May 5, 1899, Mayor penned a long letter to Agassiz explaining the scheme he had come up with for their classification. Unhappy with the system devised by the German zoologist Ernst Haeckel in his *Das System der Medusen,* published in 1879, Mayor argued that "the genera should indicate relationships firstly, and differences secondarily." By focusing on differences, observed Mayor, Haeckel had come up with many genera that "contain but a single species." As a consequence, Mayor "cut down the number of genera to a great extent" and combined many species under older generic names. Following a sound system of taxonomy and nomenclature, he revived "all old names and first spellings," which he noted for the benefit of Agassiz. Indeed, he went so far in his effort to follow the law of priority in zoological nomenclature that he did not alter names that were originally misspelled. Mayor added that in some cases he did not revive "the oldest name" if it had been "vaguely defined or published in an obscure journal . . . [and] neglected for over 50 years" because he rightly believed that such a practice would create "confusion." In addition he gave a hydromedusa the same generic name as its hydroid if both stages of the hydrozoan had been described simultaneously, but he followed the practice of allowing separate names for both stages of a hydrozoan species if the hydroid and the medusa had been described as different species. In any case, in his letter to Agassiz, Mayor noted that he had delayed any "synopses of the Trachymedusae and the Narcomedusae" until summer, when further study would enable him to correct Haeckel's "fearful tangle of confusion" about those two orders of medusae.[42]

Mayor also sketched the plan of the projected book. In his view, the work must include explanations of terms, remarks on the author's system of classification, information on "anatomy and development," and records of the geographic distribution of each species. Such information would, he declared, increase the importance of the work, as "we wish it to be one of the classics of science." The book must also include all of the species of hydromedusae, scyphomedusae, siphonophores, and ctenophores found in waters on the American East Coast, and he noted that he had completed "drawings of all but one of the southern ctenophorae." He concluded that "our system is nearer nature and farther from the stamp collector type of classification indulged in by Haeckel."[43] Just how Agassiz responded to this informative letter is unknown, but it is certain that he had to be grateful for the thorough work done by a man who was still listed only as his assistant. In fact, Agassiz had reason to wonder what he might contribute to a work that had so much promise and toward a subject on which he surely knew he had not been able to keep current. Such considerations likely account, at least in part, for his failure to push ahead with the project.

Soon thereafter Mayor received an overture from Franklin W. Hooper, director of the Brooklyn Institute of Arts and Sciences, in Brooklyn, New York, to serve as the curator of natural history for the institute's Brooklyn museum. For the offer Mayor owed thanks to his old friend Charles Davenport. In 1898 Davenport had accepted appointment as director of the summer school program of the Brooklyn Institute's Biological Laboratory, located at Cold Spring Harbor, Long Island, and a year later he left Harvard for good. In a letter to Davenport on May 13, 1898, Mayor said that if he obtained the position, he would "arrange an exhibit [that] . . . will open the eyes of the public to the fact that the study of animals is a phylosophic [*sic*] subject and not at all a mere craze for putting specimens in glass cases." He added that he would also arrange specimens in a way that would show "the chief hypotheses and laws of evolution."[44] Mayor believed that he could develop a superior museum at relatively low cost.

Quite obviously, then, he was interested in the prospective position, and he was prepared to inform Agassiz, which he would have to do by letter since he was already on his way to Loggerhead Key.

Mayor summoned the courage to do so, apologizing for the anxiety it brought to the old director. To soften the blow, he assured Agassiz that he would follow his "ideas as long as I have the honor to remain your assistant." At the same time, he suggested that he ought to remain in Cambridge and work on the book rather than go along on the forthcoming expedition, even though he would doubtless collect new species of medusae "which will *add to the book* . . . [and make it] *a complete work.*" To Harriet, Mayor confided that "probably the old man will fire me for impudence, but I don't care." Harriet replied that she was glad he had "made a stand and told Agassiz frankly you did not wish to go [on the expedition]." She encouraged him, "stick to your research, dear old man, [for] . . . [s]omething has to break sooner or later."[45] The evidence provides no information on Agassiz's response, but it appears that Mayor was wrestling over the matter of honoring or breaking his commitment.

On May 19, 1899, Mayor was back at Loggerhead Key collecting more medusae. He remained there until July 26, by which time he had decided that he would go on the expedition. Now he was worried about the possible offer of the Brooklyn Museum position, he told Harriet, because he had not informed Hooper about the South Pacific expedition. He found some solace, however, in receiving "permission" from Agassiz to write solely under his own name an arti-cle on the varying forms of a species of leptothecate that revealed evolutionary radiation. In fairness, Agassiz had no choice, for he knew nothing of the species, while Mayor had made "1,000 observations" of it. Plainly, however, Mayor still had to seek authorization from Agassiz to publish anything resulting from work done under the aegis of the MCZ. In any case, Mayor was excited, for he was certain the species would demonstrate that "when new races [of medusae] start out, they are much more variable than in the old race from which they sprang."[46] To Mayor, this was evidence of evolution by natural selection at work, a subject that held no interest for Agassiz.

Meanwhile, the prospect of yet another job came before Mayor. It would be in Honolulu, where a marine biological laboratory was to be established. Mayor would serve as curator of the radiates if the plan developed. The inquiry had come via Woodworth, but Mayor did not let his dislike of the man get in the way of a promising lead. He told Harriet she would have to agree freely to move to such a distant place and that he would follow her decision, for "you are all I live for." The climate there, he said, would doubtless be good for her constantly sore throat, a symptom, they secretly suspected, of tuberculosis. The place would be much preferable to the Brooklyn Museum, he thought, for "there is an air of quackery about the museum, which is apparent from Hooper down to the janitor." Carrying his complaint to an unwarranted extreme, Mayor called Hooper "a charlatan." He was genuinely concerned, however, over moving Harriet so far from home and worried that he might not be "independent" in the Hawaiian post.[47]

Harriet feared that Agassiz was behind the effort to get the appointment for Mayor in order to serve his own interests, and she opined that Agassiz might pay from his pocket the additional amount of salary necessary to lure Mayor to the job in Hawaii. In her judgment, Agassiz would go to any length to assure that Mayor did not back out of the South Pacific expedition, and she perceived a connection between the offer and the expedition. After all, she declared, Agassiz "has the colossal impudance [*sic*] to dispose of you as if you were his bondsman." Mayor no doubt agreed with Harriet, but he decided that he had little choice at this point. After completing his collecting work at Tortugas, he traveled to Nova Scotia in late July, hoping to obtain specimens of medusae farther north than he had found earlier. He had little success, however, and wrote to Harriet that he wondered "what 'poor' Alex will say," though vowing that he would not worry as he had done on previous trips.[48] There is always pressure, even if self-imposed, to show that a particular field trip has been productive and worthwhile. By now, however, Mayor was prepared to write about the numerous medusae he had collected in the western Atlantic and at Tortugas. He would be unable to do so, of course, until he returned from the South Pacific.

Shortly thereafter Mayor was on his way to the West Coast, arriving in San Francisco in late August 1899. On the 23rd of that month he was aboard the U.S. Fish Commission steamer *Albatross* with Agassiz and Woodworth bound for Oceania. This was to be the longest of the Agassiz expeditions to the South Pacific, and the scientific party intended to make more hauls than before. In fact, the *Albatross* made the first of its deep-sea trawls two days after setting sail. After an hour of trawling at 14,200 feet (4,328 meters), the crew brought up the net, taking three hours to do so in order to reduce chances that captured creatures would burst because of a rapid decrease in water pressure. The ship did not arrive at its first point of destination, the Marquesas Islands, until September 14. Over the next week it made its way slowly toward the Tuamotu Archipelago, and on September 27 it reached Papeete, Tahiti, in the Society Islands. After a week there Agassiz ordered the ship back to the Tuamotu Archipelago, where it made numerous trawls during the next six weeks, bringing up hundreds of specimens.[49]

In Tahiti again from November 6 to 13, Mayor resumed the collection of land snails in the genus *Partula*, an effort he had begun a month earlier, along with C. S. Kempff and Dr. H. F. Moore. Although Mayor identified Kempff as an ensign on the *Albatross,* he offered no information about Moore, but it is likely that he was Henry Frank Moore of the United States Commission on Fish and Fisheries. The party collected 735 specimens of *Partula* from six valleys on the island. In 1902 Mayor published a study of the snails in *Memoirs of the Museum of Comparative Zoölogy.* Titled "Some Species of *Partula* from Tahiti: A Study in Variation," the paper addressed the subject of species versus color varieties and the topic of geographical isolation respecting speciation among the snails of that genus. Mayor observed that A. J. Garrett's earlier analysis of *Partula* inhabiting Tahiti had resulted in some errors regarding dextral (opening on the right) and sinistral (opening on the left) characteristics of the several species. He also maintained that Garrett had in some cases "confused" varieties with species. In fact, as Mayor noted, the problem with the *Partula* in Tahiti was confounded by the great variety of colors among them. By selecting scores of specimens from six valleys differing in elevation, rainfall, and other relevant features, Mayor was able to show that "geographical isolation plays a part in the formation of new species." Mayor viewed the color variations and the openings, whether dextral or sinistral, as indications of intergrading where snails mated on the fringes of the regions (see color plate).[50]

Mayor also studied butterflies and moths, took photographs of "the tropical forest and the natives," and otherwise kept himself busy by sorting, sketching, and preserving medusae. In addition he visited the interior of islands, often in company with Moore, and, along with Woodworth and Agassiz's son Maximilian, took photographs of atolls and other coral reefs, a subject of long-standing interest to Alexander Agassiz and one that was beginning to intrigue Mayor more and more.[51] Following the argument of John Murray that barrier reefs and atolls were the consequences of submarine elevation of volcanoes and dissolution of the encircling sediment by the carbonic acid in seawater, Agassiz became more certain that he understood the nature of atoll formation. Mayor quietly held the view that the argument was wrong and that the hypothesis of geological subsidence offered by Darwin in 1842, and supported by James Dwight Dana in 1853, was a far better explanation. Agassiz did not invite Mayor to question his views on the subject.

Meanwhile the expedition continued, undertaking trawling on November 21 around the Cook Islands and four days later around Tonga. The *Albatross* sailed to Fiji on December 7 and remained in that area until the 19th of that month, moving on then to the Ellice Islands (now Tuvalu) and the Gilbert Islands. From January 17 to February 7, 1900, the expedition worked around the Marshall Islands, and then it steamed on to the Caroline Islands, Guam, and the Ladrone Islands (now Northern Marianas). The ship arrived in Yokohama, Japan, on March 4. The excursion was nearing its end, and Mayor felt that it had been relatively unsuccessful. Several weeks earlier he had complained in a letter to Davenport that the expedition "rushed from Island to Island" and therefore was "fore-doomed to be a failure from every standpoint excepting that of the study of coral reefs." He expressed disappointment that Agassiz was not concentrating more on collecting medusae, though, as the subsequent report revealed, he collected a considerable number of new species of jellyfishes. Mayor also mentioned to Davenport that he had taken many photographs, including some "representing . . . the natives of the various groups we have visited." In addition, as before, he kept a detailed journal of the expedition, recording especially his observations of the indigenous peoples of the various islands, which later served as the basis of several popular articles he published on them.[52]

Back in San Francisco on April 3, 1900, Mayor headed for the University of Chicago to present a lecture, after which he hastened on to Cambridge, where he found, in the February 1900 issue of *Psyche,* the publication of his address as president of the Cambridge Entomological Club, a position to which he had been elected early in 1899. Apparently he had managed to write the paper before he left on the expedition, for he had completed his research on the topic in June 1899. The paper was read for him at the annual meeting of the Cambridge Entomological Club on January 12, 1900. Titled "On the Mating Instinct in Moths," it was based on a study of 449 cocoons of *Callosamia promethea* collected by him and a friend in Cambridge and in Maplewood.[53]

Mayor reported that he had left the cocoons outside during the winter of 1898–99 and then carried them to Loggerhead Key on May 5, 1899. A total of 176 moths emerged between May 18 and June 8, of which 63 percent were male and 37 percent were female. Mayor was interested in determining whether the male located the female by her color or by an odor she emitted. First he placed five females in a clear jar with the opening covered by mesh cloth. Then he released five males 100 feet (30.5 meters) away. The males flew immediately to the jar. Next he placed the jar upside down in sand, which left the females visible but prevented escape of any odor. The males then flew away but returned quickly after the jar was righted. He repeated the experiment several times, with the same results. Mayor recognized, however, that he had to assume the visibility of the females through the glass. Conducting five other experiments on the same subject, in one of which he coated the antennae of males and in another he removed the wings of females, Mayor concluded that Darwin's hypothesis of sexual selection on the basis of color might not be valid.[54] The theory of evolution was becoming increasingly important to him.

The publication of the paper was certainly secondary to other developments in the lives of Alfred Mayor and Harriet Hyatt, however, with prospects of a desirable job increasing and, after four years of waiting, of being married. Mayor and Hyatt made their engagement known and set August as the date for their wedding. While he was in Chicago, Mayor had received word from the noted marine biologist Charles O. Whitman of the University of Chicago that his institution wanted to appoint Mayor to a faculty position but that he would have to wait until money was available to fund the position. Meanwhile, Franklin Hooper was pushing to get Mayor to accept the Brooklyn Museum post, telling Mayor that he was "just the man" the trustees wanted and that he would recommend an annual salary of "$2,000 for the first year with a raise later." Thanking Davenport for urging the Brooklyn trustees to hire him, Mayor said, "The biological department here [at Harvard] is literally cut in two and seems hopelessly ruined." Given that opinion about the situation at Harvard and faced with an offer that would double his salary, Mayor accepted the position of curator of natural history at the Brooklyn Museum, to begin in November 1900.[55]

As he prepared for his wedding on August 27, 1900, and the move to Brooklyn soon thereafter, Mayor continued to "work up," or classify and describe, the *Albatross* collections and to draft a paper he was invited to present at the Marine Biological Laboratory in Woods Hole, Massachusetts. For the latter he intended to offer comments on the nature of the species he had collected during his recent field trips, but he decided that he should seek the approval of Agassiz since he had collected the specimens under the aegis of the MCZ. Agassiz was not pleased, and in a letter to Mayor on June 30 he advised that "it would be preferable for you to take a different subject," saying that he planned to write on "all the material I have been collecting through you and others." Agassiz added, "it does not seem to me to be the thing for anyone to publish . . . but the head of the Expedition" lest it "lead to misunderstanding."[56] The "deep old fox" would keep control even after Mayor was gone. Moreover, the projected book on the East Coast medusae was now in limbo.

Nevertheless, Mayor would gain further recognition for himself before he left the MCZ, for the two articles he had prepared for publication solely under his own name were already in press. Both appeared in the MCZ's *Bulletin* in 1900, the first in June and the other in July. They contained descriptions of a considerable number of new taxa of medusae. In the first paper, "Descriptions of New and Little-known Medusae from the Western Atlantic," Mayor described as new to science eight species of hydromedusae, a scyphomedusa, and a ctenophore, or comb jelly—all collected between 1893 and 1898. Ultimately all proved to be valid taxa except for the scyphozoan and one of the hydrozoans. Mayor described the ctenophore, from Charleston harbor, as *Mnemiopsis mccradyi* "in honor of Professor John McCrady in recognition of his important researches upon the medusae of Charleston Harbor," where he had collected the type specimen on September 15, 1897.[57] The small comb jelly, which Mayor described as "greenish-amber in color and opalescent," belongs to the order Lobata. The article contained six plates of twenty-four exquisite figures drawn by Mayor.

In the other article published in 1900, "Some Medusae from the Tortugas, Florida," Mayor described ninety species, thirty-nine of which he believed were new to science. Of that number, sixteen turned out to be junior synonyms, having actually been described and named earlier, but a remarkable twenty-three use valid species names of hydromedusae, as is a generic taxon. Mayor had personally collected all ninety of the species during one or another of his three expeditions to Loggerhead Key during the summers of 1897, 1898, and 1899. Of the hydromedusae he collected, Mayor was especially fascinated by one he named *Pseudoclytia pentata,* though, in fact, its taxonomic status remains unclear. He believed that the animal "displays a greater tendency toward variability in various directions" than other medusae. He had seen "a great swarm" of the species on July 22, 1898, and "two such swarms in the summer of 1899." Viewing the variations as evidence of evolution, he noted that he intended to write "a special paper" on the phenomenon.[58] Such matters did not intrigue Agassiz, but he apparently raised no objection to Mayor's intention to write a paper on the subject. Mayor also observed that since not enough was known of the ctenophores of the Tortugas and the Bahamas, he could not offer a definitive statement on their geographical range. In due course he would contribute significantly to knowledge of the ctenophores, but for now he could rest content that he was one of the world's leading authorities on the medusae. He knew that Agassiz could not do without him in producing a book on the subject. That matter would have to wait, however.

# Campaigning Curator

When Alfred Mayor began his association with the Brooklyn Museum in November 1900, he had much work to do. The museum had opened only five years earlier, though it was part of a much older institution, the Brooklyn Institute. Founded in 1823, the institute flourished until shortly after the American Civil War, when it fell into a state of relative inactivity for two decades. A resurgence was under way by 1889, however, when the institute trustees appointed Franklin W. Hooper as general curator. Born in 1851 in New Hampshire, Hooper had received a bachelor's degree in biology from Harvard University in 1875, after which he studied coral reefs and algae in the Florida Keys. By 1880 Hooper had settled in Brooklyn, where he served as a science teacher and, later, a high-school principal. He became a trustee of the Brooklyn Institute in 1887. An energetic leader, Hooper succeeded in broadening the scope of the institute, which became the Brooklyn Institute of Arts and Sciences in 1890. Under his leadership the institute established a laboratory at Cold Spring Harbor, Long Island, and it created a museum in 1895. By 1904 the membership of the Brooklyn Institute had climbed to eight thousand.[1]

Although he harbored some misgivings about the long-range prospects of the Brooklyn Museum, Alfred Mayor was happy to be receiving an annual salary of two thousand dollars, twice the amount he had been paid by Agassiz. He was determined to advance the status of the institution—and, in the process, to promote his own standing as a marine zoologist. Soon after he arrived in Brooklyn, Mayor outlined to Hooper his ideas on ways to develop the Brooklyn Museum into one of the best of its kind. Rather than spend funds for whatever specimens came along, the museum must develop "a synoptic collection of those animals which best illustrate the relationships of great groups of Zoology, and this will be quite expensive as such forms are usually rare." Moreover, since he viewed the theory of evolution as vital to the study of natural history, Mayor stressed that museum collections and exhibits must reveal the "more and more complex types of life as the Earth grew older." Hence, the goal of the museum must not be to simply fill rooms with specimens but to help the public to understand important concepts in natural history. The museum must show "individuality," not merely duplicate other museums. In order to achieve these goals, he must be allowed to hire the best staff available. The costs of making the museum into an "intellectual centre" would be high, and they would slow the growth of the institution in adding to the collections. In the long run, however, the plan would pay great dividends, Mayor assured Hooper. It was already clear that Mayor envisioned a dual purpose for the museum: as a research institution and a center of learning. The latter certainly fit the motto of the museum, "For the People, By the People," but the former, while obviously an aim of the institution, seems to have been the major goal of Mayor, who was influenced by the character of Harvard's Museum of Comparative Zoology, with which he had been associated for eight years. No doubt Mayor valued the museum's educational function, but scientific research held dominance in his mind.[2]

Not long after outlining his ideas for the museum, Mayor sent to Hooper a letter he had received from Davenport endorsing another part of Mayor's plan, namely, a focus on building collections of "local Zoology" rather than of specimens from exotic regions, for the latter, said Mayor, tend to reinforce the notion that a museum is merely "a *curiosity* shop!" In addition Mayor also reiterated his request for establishing a research journal under the auspices of the museum. It appears that Hooper had neither anticipated such major requests from the new curator nor expected to have to ask the trustees for additional funds. In early December, Mayor sought approval to purchase a collection of fossils and a mounted skeleton of *Bos urus,* the extinct wild ox of Europe. Hooper informed him, however, that funds had been reduced not only for such purposes but also for supplies, including alcohol necessary for preserving many specimens. Disappointed over this sudden setback, Mayor expressed regret, adding, "I think that it would now be most unwise to plan any collecting expedition to the Bahamas in the early spring." He was determined, however, to move forward with his plan to "obtain a very complete . . . collection of the animals and plants of Long Island," and he suggested names of people who could collect vertebrate specimens in the region. In addition he would personally collect marine invertebrates from the shores and waters of Long Island. Mayor also intended to spend time at the Cold Spring Harbor Laboratory conducting experiments with insects. Of course, Mayor was interested in the work of his friend Davenport at that laboratory, and it was there that he became a friend of Eugene G. Blackford, president of the laboratory's board of managers. Mayor honored Blackford in 1910 by naming a new genus of leptothecate medusae after him (*Blackfordia*), and two species established by Mayor, *Blackfordia manhattensis* and *Blackfordia virginica*, continue to bear his name.[3]

In Mayor's scheme of programs for the museum, public education on topics in natural history would play a key role. Thus, on December 10 he recommended to Hooper that the museum hire Louise K. Miller of Dayton, Ohio, to develop that facet of his plan. In his interview with Miller, Mayor was impressed by her "general intelligence, frankness, and charm of manners." He wanted Miller to present one lecture each day to a group of schoolchildren and one lecture per week to the general public. In addition she would develop an illustrated serial publication "devoted to popularizing science." Mayor also worked to establish a series of illustrated lectures in zoology by prominent authorities, and he was quite successful in doing so. During his first year at the museum he managed to line up two highly distinguished scientists: Thomas Hunt Morgan, of Bryn Mawr College, for a series of lectures on regeneration and embryology; and Addison E. Verrill, of Yale University, for talks on comparative zoology. In addition he personally presented a ten-lecture series on the radiates, which included jellyfishes, corals, anemones, sea stars, echinoderms, and other invertebrates characterized by radial symmetry. He also gave an eight-lecture series and conducted eight "field meetings" on insects. During the following three years he continued the series, getting Verrill to return on two occasions and bringing the outstanding biologist Edmund Beecher Wilson to the museum to speak on the subject of how cells divide and on the laws of heredity. Mayor, too, participated actively in presenting several series of lectures—for example, five lectures on "The Deep Blue Sea and Its Life," ten on "Marine Animals," six on "The Faunae of the North Atlantic Coast," eight on "The Evolution of Animal Life," and several others, including ones on coral-reef formation, "The Gulf Stream and Its Life," and "The Beaches of Long Island and Their Life."[4] No doubt, Mayor could have persuaded other distinguished scholars to present lectures at the museum, but lack of funds kept him from doing so.

Hooper had his own aims and vision for the Brooklyn Museum, however, and early in January 1901 he suggested that Mayor should run "a summer school in Natural History at the Museum" for schoolteachers. Mayor was not opposed to the school, but devoting his summers to it would curtail his collecting activities for the museum and force him to give up his own research. He could not build the collections of local and state fauna and flora, he informed Hooper, if he had to run the summer school. Such an assignment, he bluntly told Hooper, would come "at the expense of the broader interests and ultimate usefulness of the Museum as an educational factor," for the summer-school participants would need specimens for dissection since illustrations alone would be insufficient. Moreover, he said that he must have every summer for research. In a letter dated January 21, 1901, Mayor told Hooper that he was unwilling to run the summer school and suggested that the director assign the task to Louise Miller. It appears, however, that the museum did not hire Miller after all, the reason for which is unknown. The decision not to employ her could have been due to budgetary concerns, although in his reply to Mayor on January 22, 1901, Hooper referred to his intention to increase "the Museum staff."[5]

Hooper was not happy with Mayor's response. Noting that the museum was committed to the summer school, he said that "all officers of the Museum" were obliged to participate in it. He maintained that Louis Agassiz had helped to educate numerous teachers via his summer programs, and he stressed his commitment to "the teaching power of the Museum." Then, in a statement that likely irritated Mayor, Hooper added, "It is the personal influence of men connected with the Museum rather than the collections that are on exhibition, that will be of lasting and important value." He ended his letter by saying that he was sure Mayor would want to help in developing the summer school.[6] Mayor relented, thus experiencing another setback to his plan. He continued to press for development of the programs he favored most, however.

Mayor aimed to convince Hooper that he knew what was best for the museum. If the museum was "to rise above the level of insignificance," it must focus on building its collections, and that would require dependence on trained naturalists, not on donations alone. The museum has "a corps of Curators" who can do the job, said Mayor, but each of them would need "a maximum vacation of three months per year" for the activity. Mayor argued that his approach would cost less than the purchase of specimens, and he told Hooper that "one need only inspect our local collections to know how small and imperfect they are in every respect." If the museum failed to follow his plan, he added, it would have no prospect "of any real success for years to come."[7]

A few weeks later Mayor again emphasized the importance of building collections of local fauna and flora and also of developing an educational exhibit that revealed "the life histories of Long Island plants and animals." He reminded Hooper that it would not be a good idea to keep on "expanding in many directions when we are really doing nothing well in any direction." After stating many other needs, Mayor repeated a point he had already stressed: "It should be remembered that a Museum is not a building full of show cases. We must have a library and a laboratory for research and instruction." He also told Hooper that museums in Europe enjoyed esteem "because new ideas emanate from them and science is advanced by their efforts. . . . Let us give the people *ideals,* rather than *sensations.*" All of this, he reminded Hooper, would require a first-rate staff and a considerable amount of money.[8] Clearly, Mayor's notion of the role of a museum differed from that embraced by Hooper, and the latter devised a scheme that he thought would bring Mayor into line.

On May 25, 1901, Hooper informed Mayor of a plan for reorganization, by which the latter would become curator of zoology. Mayor replied a few days later, saying that he wished to remind Hooper that he had been "officially appointed as 'Curator in the Natural History Sciences'" and would not "be willing to serve under a title implying lessened responsibilities." He also suggested that the proposed title was tantamount to "declaring lack of confidence" in his work. If the trustees exercised their right to change his title and duties, he would be compelled to resign, he said. Hooper responded immediately, contending that the proposed title was in no way intended to "render less valuable your relation to the Museum," and he declared that he saw "much more dignity and value" in a title associated with "a well defined subject." Ironically, he told Mayor that he should view the new title as "a promotion." Eventually the museum would have many curators in the natural sciences, with each designated as a curator in his specialty, suggested Hooper, without acknowledging that he was agreeing with Mayor about a lessening of his authority as curator in charge of all the natural sciences. Hooper backed away from the change, however, saying that the matter should be deferred until the museum got into a larger building.[9]

"You ought to have seen the old Hypocrite whine . . . and utter . . . no end of insane flatteries," Mayor told Harriet, adding, "It is highly probable that the old fox realizes that the Trustees would not support him" and that he had really hoped "I would voluntarily yield" to the change. It is likely that Mayor was right, for he was apparently held in esteem by the trustees. He took advantage of the retreat by Hooper, who in his view told enough lies "to make a cat giggle."[10] Whether or not Hooper was hypocritical cannot be judged from Mayor's contention that he was, but it seems obvious that Hooper wanted to keep the curator in check. By backing away, however, he gave Mayor an opening to ask him to recommend to the trustees that his salary be paid over nine months and that he be given the entire summer for research. For Mayor, research and publishing were essential to professional development. After all, he had been trained in a museum designed mainly for research, not primarily for exhibits of its collections or principally as an institution for education. The difficulties between Mayor and Hooper might have been avoided, or at least reduced, if Hooper had been candid when negotiating with Mayor and if Mayor had made his intentions clear then. However, each man seemed to be operating on the assumption that the other understood his notion of the primary purpose of the museum, the roles of its staff, and the financial resources available to run the institution.

Even before the unhappy encounter, Mayor had privately complained that the trustees had viewed his intention to do "statistical research on color patterns" in several species of the museum's huge collection of Lepidoptera as "an 'extravagant' waste of money." He also grumbled over being forced to mingle specimens of lepidopterans collected for research with those sold by the museum. Despite those views, however, the trustees readily approved Mayor's proposal to establish a research journal, to be called *Science Bulletin,* under the aegis of the museum. Yet they probably did not understand what Mayor had in mind, likely viewing the publication as a form of good publicity rather than a strictly academic one. For Mayor, however, the *Science Bulletin* was to be a forum for "nothing but original work of a high character." He was encouraged, he informed Davenport in a letter of March 10, 1901, that Hooper had assured him that "no paper will be published which meets with my disapproval." In fact, he told Davenport that he would "rather resign than see the Bulletin made use of for advertising or for the publication of poorly executed work." He had worked hard, he added, to win approval of the *Science Bulletin,* and he swore to make it succeed or resign if he failed. Two months later Mayor expressed to Davenport his intention to publish "a series of 'Memoirs of Art and Archaeology'" and "a 'Journal' devoted to popular articles, reports of lectures, and . . . work by the various departments." Mayor concluded by stating that his purpose was "to give the Museum a standing in the world," and he was unequivocally committed to excellence.[11]

To assure a standard of high quality in the first issue of the *Science Bulletin* and to get the new journal launched within two months after the trustees gave him the nod of approval, Mayor published one of his own studies, "The Variations of a Newly-Arisen Species of Medusa." It was a paper he had received permission from Agassiz to publish but had not completed earlier because of his busy schedule. In the article Mayor included sixty-seven well-executed drawings of variations in the lips and in the radial canals of one thousand specimens of the leptothecate medusa *Pseudoclytia pentata,* from the Dry Tortugas. Employing statistical methods, he found a significant difference that led him to conclude that *P. pentata* "departed widely from the fundamental type of all other Hydromedusae" and that it had been derived from some species of *Epenthesis.* He concluded that the "newly-arisen medusa" was evidence of evolution at work. Ultimately, however, his study came into question, as the species is possibly the same as *Clytia folleata* (McCrady, 1859) and the variations are typical of that species. We believe, however, "that further material is necessary to resolve the identity of *C. folleata.*"[12] In any case, Mayor's approach to the inquiry represented a commendable effort to apply statistics to the study of a biological subject. Mayor also published papers he received from others as well as more of his own in the next few issues of the *Science Bulletin*. He had a long way to go, however, in lifting the Brooklyn Museum publication to the rank of the best national or international scientific journals, or to that of major museums in the United States or abroad.

More important to the reputation of Mayor was the publication of descriptions of medusae collected during the expedition to the South Pacific in 1899–1900. Appearing in the *Memoirs of the Museum of Comparative Zoölogy, at Harvard College*, the article—with Agassiz's name listed first and Mayor's second—offered descriptions of twenty hydromedusae, eight scyphomedusae, ten siphonophores, and four ctenophores. Of the eleven Agassiz-Mayor taxa of hydrozoans, six are currently valid, while one of their four taxa of scyphozoans and one of their two taxa of ctenophores are presently valid. One of the species the authors believed to be new was the coronate scyphomedusa *Nausithoe picta*. In describing it Mayor wrote: "The bell is translucent and milky in color. The ocelli and gonads are of a rich brown color, and the entodermal core of gastric cirri are [*sic*] dark blue." A colored drawing of the species he later prepared reflects the notable beauty of this creature. The ctenophores, or comb jellies, were *Pleurobrachia ochracea* (=*Hormiphora ochracea*) and *Lampetia fusiformis* (=*Lampea fusiformis*), each in the order Cydippida.[13]

Meanwhile, soon after the Carnegie Institution of Washington was founded in January 1902, Mayor applied to it for a grant to assist him in completing a study of land snails in the family Achatinellidae begun by Alpheus Hyatt. However, the reply, received many months later, was negative. Clearly dissatisfied with the Brooklyn job by now, a discontented Mayor was casting about for another position. "I could never love Brooklyn," he confided to Harriet in January 1902, "even if we had a fortune here each year."[14] A few weeks later, discouraged that the Brooklyn Institute trustees had declined his request to pay his salary over ten months (altered from his earlier request of nine months) and allow him two months free for research, Mayor told Harriet that he would soon inform Hooper that the decision "will oblige me to leave here as soon as I can obtain another position." At this point he believed that he might be appointed as curator of the museum of the Boston Society of Natural History and said he was "beginning to ache for it," though he believed it was less promising than the Brooklyn position. He expected the Boston Society to offer a yearly salary of at least twenty-five hundred dollars and to give him three months of research leave each year. Nothing came of his hope, but in the meantime an event had occurred in Washington, D.C., that would affect the course of his life.

On January 29, 1902, the newly appointed board of trustees of the Carnegie Institution of Washington (CIW) accepted from the wealthy entrepreneur and philanthropist Andrew Carnegie a deed establishing a trust to be used for promoting research. Among the numerous advisory committees soon appointed by the trustees was the Committee on Zoology, whose members were Henry F. Osborn, C. Hart Merriam, Alexander Agassiz, William K. Brooks, and Edmund B. Wilson. The latter three members worked in marine zoology, and as it so happened, each was well acquainted with the work of Alfred G. Mayor, who, long before the committee made its report, had begun to champion the establishment of a marine biological station at Tortugas. In fact, the Committee on Zoology was supposed to consider a range of possibilities, but given that three of its members were greatly interested in marine zoology and that a fourth, Osborn, was, like Brooks and Wilson, a trustee of the Marine Biological Laboratory, some biologists were rather certain that the committee would favor marine studies. If Mayor had any doubts about the committee's recommendations, he never revealed them. Indeed, by April he was working hard to assure not only that the Carnegie Institution would establish at least one marine biological station but also that it would be placed on Loggerhead Key, Dry Tortugas. On this matter he conferred often with Charles Davenport, who had his own interest in the Carnegie Institution's decision. Recognizing the thrust of the Carnegie trustees toward collective endeavors rather than individual grants, Mayor believed that he could readily demonstrate how a marine laboratory would fulfill their expectations.[15]

Devoted to the nation to which he had immigrated and in which he had ultimately prospered, Andrew Carnegie had decided late in life to endow a research institution that would move the United States to the forefront in science. New discoveries intrigued him, and he believed that "exceptional men" would find such discoveries if they received outstanding support through the Carnegie Institution of Washington. They would engage in collective efforts rather than work in isolation, however, for Carnegie believed that the day of the rugged individualist such as himself had passed. Science stood in need of centralized structuring. Although especially fascinated by celestial discoveries, Carnegie allowed the trustees of the new institution to decide which areas of science they would support. Carnegie would serve as a patron of science, but he would leave the choice of departments to those who knew best what was needed to move the United States to a commanding position in science.[16] Alfred Mayor recognized that marine biology was one of the fields that could readily fulfill the desire for discoveries, and he set his mind on persuading the Carnegie Institution of Washington trustees of the importance of creating a biological station that would emphasize tropical marine life.

Marine biological stations had begun to flourish in Europe long before this time, and efforts to establish similar laboratories were under way in the United States by the 1870s. As the American zoologist Charles A. Kofoid observed in

1910 in a survey of biological laboratories in Europe, "the great problems in biology" raised by Charles Darwin's theory of evolution not only helped to entrench the place of the biological laboratory in colleges and universities but also served to boost the importance of studying "at the seashore." Among the most prominent of the marine biological laboratories in Europe when Mayor first sought to establish a station in the Dry Tortugas were the Stazione Zoologica in Naples, Italy, and The Laboratory, Marine Biological Association of the United Kingdom, Plymouth, in England. The station in Naples was founded by Anton Dohrn in 1872, and the laboratory in Plymouth was established in 1888 by the Marine Biological Association of the United Kingdom.[17]

Meanwhile, in the United States, Louis Agassiz had operated a marine biological laboratory on Penikese Island, near Cape Cod, during the summer of 1873. Agassiz died in December of the same year, but his son Alexander, recognizing the importance of marine biological laboratories, continued the effort, establishing one in his home in Newport, Rhode Island, in 1875. Named the Newport Marine Laboratory, the facility was moved into a new building on the Agassiz property in 1877. It was there, in the summer of 1892, that Alfred Mayor received his introduction to marine biology.[18]

Other marine biological laboratories were also springing up in the United States around the same time, including one located in Salem, Massachusetts. This laboratory was established by Alpheus Packard through the auspices of the Peabody Academy of Sciences and operated during the summers of 1876–81. In 1878 the Johns Hopkins University zoologist William Keith Brooks organized the Chesapeake Zoological Laboratory, initially located at Fort Wool in Hampton Roads, Virginia. Soon thereafter Alpheus Hyatt instituted a summer seaside laboratory at his home in Annisquam on Cape Ann and ran it until 1886. Hyatt's laboratory was moved to Woods Hole, Massachusetts, in 1888 and became the Marine Biological Laboratory (MBL).[19]

At the time that Mayor was proposing the opening of a marine biological laboratory in the Dry Tortugas, the MBL had already developed a sound reputation for its work, albeit not without concern for a more solid foundation of financial support. As Mayor was contemplating a campaign to generate interest in establishing a tropical marine biological laboratory, his former Harvard University mentor Edward L. Mark was working to establish a marine biological station in Bermuda. Mark succeeded in 1903, when the Bermuda Biological Station for Research was opened. Mayor said little about the laboratory in Bermuda, and it posed no direct challenge to his plan for a tropical station. Located in a subtropical climate, with winter water temperatures dropping as low as 61–63°F (16–17°C), the Bermuda station, as noted by C. M. Yonge, was "marginally tropical."[20] Moreover, it did not enjoy the immediate prospect of the sizable funding Mayor hoped to obtain from the CIW.

Mayor's friend Charles Davenport was apparently aware of the potential problems of establishing a laboratory in the remote Tortugas and suggested early on to Mayor that Biscayne Bay, Florida, might be more suitable. Mayor expressed concern, however, that the suggested area had "a very limited fauna within reach of it" and "no coral reef[s] to speak of." Work during summers would be difficult there also, added Mayor, "on account of mosquitoes and heat." Despite "its relative inaccessability [*sic*] there is no place on the Florida coast that approaches the Tortugas for the variety of life," he concluded. First, of course, he had to be sure that the Carnegie Institution would in fact choose marine biology as one of the areas it would fund. Thus, on April 7 he wrote a letter to the institution's president saying that he supported "a research laboratory for the study of marine zoology of the tropical Atlantic." The laboratory should be able to accommodate as many as twelve zoologists at any given time, he said. Then he touted the Tortugas as the best place for the lab, noting that the small islands were surrounded by coral reefs and "pure deep ocean water" and were situated close to the Gulf Stream, which has "vast numbers of pelagic [open ocean] animals." He also stressed that the Tortugas keys were "healthful" since mosquitoes rarely inhabited the area.[21]

To ease the minds of institution trustees over the costs of establishing and maintaining a laboratory in such an isolated place, Mayor said that only two portable buildings would be necessary, one for the laboratory and the other for sleeping quarters. In addition the researchers would need rowboats, a sailboat, and a fifty-foot yacht powered by a twenty-horsepower naphtha engine. Other costs would include travel expenses for the researchers, equipment, and supplies. In order to make his proposal more appealing, he stated that while many excellent biologists worked in the United States, they had done "little creditable research upon the fauna of the tropical Atlantic." He also observed that the United States did not compare favorably with Norway and Italy in marine research. Mayor stated that to lead the laboratory he was proposing, the director must be someone experienced "in marine ecology" and "capable of carrying out research of a high order." That he was thinking of himself for the job became even more obvious when he added that the director must also be able to command the station's yacht.[22]

Apparently, Mayor had already informed Hooper that he was urging the Carnegie Institution to establish a marine research station at Tortugas, but somewhat disingenuously he said, "I doubt that I would desire to be the director" if his proposal were accepted. Meanwhile, Hooper saw an opportunity to associate the Brooklyn Institute with any marine biological laboratory sponsored by the Carnegie Institution, although he was not especially interested in Tortugas. Instead, he favored Cold Spring Harbor on Long Island Sound, where the Brooklyn Institute already operated a laboratory, and in mid-April he urged Mayor to go to Washington to talk with Carnegie Institution trustees about the possibility of doing so. Mayor declined, saying that "the time is not yet ripe for such a move." He believed that Agassiz would "probably have much control over the matter" and would prefer Tortugas. Moreover, Mayor was aware that his friend Charles Davenport was working up a proposal to the CIW for a biological laboratory at Cold Spring Harbor that specialized in experimental research on evolution. Hooper then pressed Mayor to talk with Agassiz and urge him to support Davenport's proposal. Understandably, Mayor again declined, arguing that it would be best to wait and see whether the Carnegie trustees would decide to support the Cold Spring Harbor laboratory, but he told Hooper that assurance of cooperation by the Brooklyn Institute would likely help to persuade the Carnegie trustees to approve the proposal.[23]

Finally acceding to Hooper's request, but no doubt wishing to support Davenport's proposal and to see what he might learn about support for the idea of a Tortugas laboratory, Mayor went to the MCZ to see Agassiz. He learned little, and on May 2 he informed Hooper that Agassiz said he had "at present . . . nothing to say concerning marine laboratories." A few days later he wrote another letter to Hooper, who was by then likely wondering about Mayor's intentions. Mayor told Hooper that, in his recommendation to the Carnegie Institution for a laboratory at Tortugas, he had not proposed himself as director of it but, he added, "if the Carnegie Institution chooses voluntarily to offer me the Directorship . . . I think I would be pleased to take it." To reduce the possibility of a breech with Hooper and the Brooklyn Institute trustees, however, he claimed that the Brooklyn Museum would benefit, for he would arrange to place "all of the unique specimens . . . [in] this Museum, or [in] any other institution with which we might have the honor to be connected," the latter part of the statement suggesting that Hooper should keep him associated with the Brooklyn Museum if the Tortugas plan was accepted.[24]

On May 22, 1902, Mayor received a letter from Agassiz informing him that "the [applications] to the Carnegie Trustees have become so numerous that we are compelled to ask for written statements of requests." A day later Mayor told Hooper that Agassiz's letter made him believe that Agassiz "has already been consulted . . . concerning the use of funds by the Carnegie Institution." Mayor indeed perceived the situation correctly, but he was unaware, or so it seems, of a major proposal facing the Carnegie trustees. Agassiz also informed Mayor that he was not a member of the Committee on Zoology, but Mayor told Harriet that he thought "Alex is whooping." In fact, Agassiz was not a member of the committee, as he had recently resigned from it, though he made no effort to clarify the matter for Mayor, nor did he tell Mayor that he hoped for a grant from the Carnegie Institution for another of his own expeditions to the Pacific.[25]

Unbeknownst to Mayor, his proposal was receiving no serious attention from the Carnegie trustees, for they were engrossed in the prospect of acquiring the Marine Biological Laboratory located at Woods Hole, Massachusetts. The MBL, founded in 1888, was becoming an important American institution for marine biology, but as it was encountering financial difficulties, its trustees proposed that the Carnegie Institution support it by a grant. On March 11, 1902, the institution's Executive Committee appointed one of its own members, John S. Billings, a physician and director of the New York Public Library, to explore the matter. Billings endorsed the application for a grant, but he recommended that the Carnegie Institution make a permanent arrangement with the MBL that would necessitate financial control of the laboratory. Eventually the MBL trustees proposed joint control, but the Carnegie Institution trustees backed away, desiring to avoid a situation over which they had to share authority. By mid-July 1902 it appeared that the MBL trustees might transfer control of the laboratory to the Carnegie Institution. The matter was not settled until late October, however, when it became clear that the MBL would not forfeit its independence.[26] At last the Carnegie trustees were ready to entertain other proposals.

Meanwhile, Mayor had kept himself busy. From June 23 until near the end of July he was doing research at Loggerhead Key, mainly on the palolo, a segmented worm in the phylum Annelida. He had already managed to complete an article titled "Effects of Natural Selection and Race-Tendency upon the Color-Patterns of Lepidoptera," which he published in the Brooklyn Museum's *Science Bulletin* in 1902. It was an elaborate study containing thirty tables that showed numbers of spots, bands, and combinations of the two in the forewings and hindwings of the species he examined. Mayor concluded that evolution among lepidopterans was not due to natural selection but to "persistent hereditary tendencies."[27] How the latter might be related to the former seems to have escaped him, as it did many of his contemporaries who had become fascinated by the subject of heredity. Of course, the study of the species of *Partula* conducted by Mayor in 1899 had addressed the subject of variation and development of new characteristics. Darwin's theory of evolution continued to be an important topic for Mayor.

In mid-September 1902, as he was exploring the prospects of leaving the Brooklyn Museum and while he was waiting, with less hope, for good news from the Carnegie Institution, Mayor received an invitation to write descriptions of medusae of the Hawaiian Islands obtained on a U.S. Fish Commission expedition in 1902. He accepted the offer on the condition that the commission would publish figures of any new species he described. He soon completed the project, but it did not appear in print until 1906. The collection was small, and the specimens had been preserved in formalin, which resulted in the destruction of the statocysts, or balancing structures, in the hydromedusae, thus "rendering specific identification difficult or impossible." Mayor described the seven better specimens, of which he thought two were new: a cubozoan and a hydromedusa. He rendered superb, detailed drawings of those two, but neither taxon proved to be valid.[28] In any case, others recognized Mayor as an able person to turn to for detailed scientific descriptions of medusae.

By December 1902 Mayor was aware that the prospect of Carnegie Institution support of a new marine biological laboratory was once more open, and he wasted no time in resuming his effort to champion Loggerhead Key as the ideal site for such a station. Early in January 1903 he drafted an article for publication in *Science*. "The Tortugas, Florida," he said, "probably surpasses any other situation in the tropical Atlantic, in the richness of its marine fauna and in natural advantages for the study of tropical life." Noting that Louis Agassiz, Alexander Agassiz, and others, including Charles C. Nutting, a University of Iowa professor and an expert on hydroids, had made expeditions to the area, he observed that the Brooklyn Museum had sponsored research there only a few months earlier, which was, of course, a reference to his own work. The Tortugas, he said, consisted of "seven low, sandy islands and numerous reef flats . . . [that] partially enclose a lagoon about ten miles long and six miles wide [sixteen by ten kilometers]," and he noted that the islands were surrounded by "pure, deep ocean water" and lay close to the Gulf Stream. Playing down the remoteness of the islands, he pointed out that two of them were inhabited: Garden Key, occupied by Fort Jefferson, by then a naval base; and Loggerhead Key, the site of a U.S. lighthouse. Although he admitted that heat and humidity ran high during the summer days in the Tortugas, he emphasized that "the nights are cool" and that a steady breeze blew during the daytime, making it "possible to retain normal health and energy."[29] Mayor said nothing of the potentially devastating hurricanes that could sweep across the islands.

Aware that he needed to persuade not only the Carnegie Institution trustees but also fellow biologists of the benefits of a marine station in the remote Tortugas, Mayor stressed the opportunities for research there. In a single month of collecting, he had, for example, collected 265 species of marine invertebrates at a depth of less than one fathom (1.8 meters). In fact, asserted Mayor, the hauls he had made were richer than those in Bermuda, and he observed that several species of birds and turtles, especially the loggerhead turtle, "now becoming rare," can be found on all of the islands. Particularly important, he noted, are the coral reefs of the region, though he observed that the Scleractinia, or stony corals, "are poorly represented" because of "a dark-colored water" that had drifted from the mainland to the reefs in 1878 and killed "great numbers" of them. Yet, based on their size at the time he wrote, three species of stony corals had "survived in considerable numbers": *Porites* (presumably *P. porites*); *Orbicella,* now *Montastrea* (either *M. annularis* or *M. cavernosa*); and *Meandrina* (certainly *M. meandrites,* as it is the only species of *Meandrina* in the region). Thus, declared Mayor, "we have at our very door a tropical fauna far surpassing in richness that of Naples," and to whet his challenge, he stated that "the great monographs of the Naples Laboratory should be an incentive to do even more and better things." In fact, as the historian Jane Maienschein has stated, the Statione Zoologica had become a "biological mecca." Drawing biologists from far and wide, it was well equipped, had a fine library, and produced excellent publications. Mayor hoped to emulate the Naples Station in developing the Carnegie Institution's new department. As he had done in his original proposal to the Carnegie trustees, Mayor listed the facilities and equipment needed for a laboratory on Loggerhead Key.[30]

As a second step in his campaign to influence the Carnegie trustees and to drum up interest among biologists, Mayor sent letters to eighty of the most prominent of the latter, asking their views on locating a laboratory at Tortugas. To the Columbia University zoologist M. A. Bigelow he wrote, "I think the subject [of a research station at Tortugas] should be agitated and hope that we may enlist the sympathies of the Carnegie Institution." By late February, Mayor had received responses from three dozen of the biologists, and six others sent replies over the next several weeks. All of the respondents favored the establishment of a marine laboratory, but some were equivocal about Tortugas as the ideal place. Among those who held few or no reservations were Frank M. Chapman, William H. Dall, Frank R. Lillie, David Starr Jordan, Thomas H. Montgomery, Charles C. Nutting, and Alexander Agassiz. The noted malacologist Dall said, "I heartily approve the proposition to establish a biological station at the Tortugas," but he thought that relative inaccessibility might be a drawback. The American Museum of Natural History ornithologist Chapman favored the plan, indicating that it would provide an opportunity to study colonial nesting birds and migration of other avian species. In fact, Chapman filled four pages of his letter with comments on types of ornithological studies that could be conducted on the Tortugas islands.[31]

Not unexpectedly, Davenport also wrote in support of Mayor's plan, stressing that the fauna of the region was not well known and touting the location because it was free of "parasitic diseases." Lillie said, "I highly approve of the plan," but he viewed the laboratory as only one of a series of stations needed along the Atlantic coast. As one who had traveled to Dry Tortugas and collected specimens there, Nutting endorsed Mayor's proposal, stating that he believed researchers in America's western and southern universities and colleges would especially welcome an opportunity to work at Tortugas. Support of the proposal by Jordan was particularly welcome because of his status as a renowned ichthyologist and Stanford University president. Jordan said that the Tortugas location would be ideal for the study of coral reefs and the animals associated with them. Montgomery viewed Tortugas as "a splendid place" for the study of polychaetes (segmented worms), nudibranchs (sea slugs), copepods (crustaceans), and siphonophores (hydrozoans such as the Portuguese man-of-war). Montgomery stressed the need for studying those and other marine species "as a whole and in their normal environmental conditions."[32]

Perhaps no endorsement meant more, however, than the one from Agassiz. Of the Tortugas, he said, "there is no locality like it." Since Mayor had not solicited a letter from Agassiz, he was especially pleased to have this resounding endorsement. Seizing an opportunity, Mayor asked Agassiz to send a letter to the Carnegie trustees expressing support of a research station in the tropics. Of course, Mayor knew that Agassiz favored Tortugas rather than Nassau, Bahama Islands, but in order to appear fair he said he would assess Nassau as a site during the coming summer, when he would be collecting specimens there. He had previously told Davenport the same thing, but while he probably intended to give equal attention to Nassau, he could not subdue his clear preference for Tortugas.[33]

Four of the respondents were less enthusiastic about the Tortugas site. All of them were noted zoologists: Edwin G. Conklin, Thomas Hunt Morgan, Addison E. Verrill, and Charles O. Whitman. Conklin simply noted that neither he nor his students would be likely to use it because of its inaccessibility and the absence of terrestrial fauna. He suggested that Jamaica might be a better place. Jamaica was also the preference of Morgan, mainly, he said, because of the remote location of Tortugas, though he believed that stipends for the researchers' expenses might help to offset the problem. Among the last to respond to Mayor's query was Whitman, who supported the Tortugas site but, in effect, offered no encouragement. In Whitman's opinion, the great need was "a single station which will draw naturalists together into working contact." The creation of more stations, he said, would "scatter forces and interests." Although the Tortugas region abounded in fauna and flora, he said, "what we need most of all is not new material but concentration of our forces and attention on material that lies nearest to hand." He declared that "a really good station has yet to be created" and asked whether it was wise to "multiply . . . weak stations." If Mayor's proposed laboratory contributed toward the goal of the strong station he envisioned, he would be in favor of it, but if "the rambling race for novelties" continued, marine science would not advance.[34]

By referring to "novelties," Whitman was expressing a concern that Mayor's marine station would concentrate on searching for new species and ignore developing areas of experimental biology. His fear was unfounded, but after all, much of Mayor's work at that point had been done in the field of systematics. On May 6, 1903, Mayor wrote to Whitman that his views should be published in *Science*. He agreed that a central station was desirable and said that the MBL should be that station. Yet, given the long coastlines of the United States, he argued, zoologists have unlimited opportunities for marine research. The MBL knows nothing of the fauna of the tropical region of the Atlantic, added Mayor, and thus, "special research stations would aid Woods Holl [the MBL] and increase its prestige." Mayor had already heard from Davenport that Whitman was "cool" toward the idea of a tropical laboratory, but he believed that the thirty-four favorable responses he had received when he wrote to Davenport on March 2 revealed strong support for his proposal. He stated that he would welcome Whitman's ideas, but since Whitman wielded considerable influence in the zoological community, he was no doubt worried.[35]

In the meantime, however, Mayor was continuing to promote the idea of a tropical laboratory. In the April 24 issue of *Science* he had once again noted that the "systematic study and classification of forms in our tropical waters is glaringly incomplete." The emphasis on systematics was, of course, what Whitman had criticized, but Mayor was not simply an old-fashioned naturalist who viewed the primary goal of biological research as descriptive taxonomy. Quite rightly, he considered systematics to be important, but this time he also noted that a tropical marine station would allow researchers to "rear larvae or carry out physiological experiments," and he argued that these studies could be done more successfully in the natural habitat of the tropical species. After indicating that no university had been able to establish a permanent marine biological laboratory in the tropics, Mayor stated that such was now possible through the Carnegie Institution. Then, to clinch his case, he summarized the responses he had received from leading zoologists, quoting at length from ten of the letters.[36]

His case was strengthened soon thereafter by a letter from Davenport to the editor of *Science*. Davenport declared that two tendencies in biological research were developing: (1) "to investigate the phenomena of structure, development, and function in the individual"; and (2) "to consider individuals, in mass or as species, as form-units bearing the imprint of environment and adapted thereto and as constituents of faunas." For students of the first, argued Davenport, "one large central laboratory" would be necessary, and for students of the second, "several laboratories, widely separated, in diverse environments" were essential.

The MBL would be ideal for the first, he declared, while a laboratory at Cold Spring Harbor, one at Beaufort, North Carolina, another in Bermuda, and a fourth in the Gulf of Mexico or in the Caribbean Sea would meet the needs of the second. He offered a similar scheme for the Pacific coast of the United States. Of course, Davenport was pushing for his own Carnegie-sponsored laboratory at Cold Spring Harbor, but he was also endorsing the plan of his friend Mayor. By noting the "magnificent chain of Biological research stations" in Europe, Davenport played to the pride of his countrymen. American universities, he argued, were too far removed to assume the task, but the Carnegie Institution could establish one at Tortugas.[37] Mayor could not have hoped for stronger support, but he would be compelled to wait to see whether his campaign would bring the results he wanted. He continued to insist that he was not vying for the directorship of the proposed laboratory, but his actions suggest otherwise.

Another article that Mayor wrote for *Science* appeared in the May 29, 1903, issue. Titled "The Status of Public Museums in the United States," the article recommended improvements in the nation's museums. According to Mayor, America's libraries had been steadily improving, but the country's museums were, on the whole, lagging behind. In fact, he contended, many university-owned museums of natural history were little more than storage places for specimens and were thus not research oriented. As he viewed the matter, American universities ought to follow the pattern that existed in Germany by turning their museums of natural history into centers "for the intellectual life of the graduate student in the natural sciences, [and] the curators should be his teachers." In addition, argued Mayor, more museums "devoted to the industrial arts, . . . historical exhibits," and the arts should be established. It is clear, however, that natural history museums were his main concern, a point he stressed in a letter to Richard Rathbun, of the United States National Museum. Rathbun had expressed interest in Mayor's article and wanted to know more about his ideas, especially on the role of curators in natural history museums. Mayor replied that he was opposed to the idea of the curator as a mere "care-taker" of collections and expressed the view that the curator "should increase knowledge by his researches and study." While the collections management duties of the curator were necessary, they were "not of paramount importance in comparison with [his] higher duties," and, added Mayor, the United States National Museum should set an example in that way.[38]

The reason Mayor chose to offer his ideas on the nature of museums at this point is not known, but he may have intended to open up the possibility of being invited by a university to direct its museum of natural history, especially since he was uncertain about the success of the proposal lying before the

Carnegie trustees. Another possibility was that he may have hoped to make more of the Brooklyn Museum, as suggested by his complaint to Hooper that Brooklynites were not giving enough money to it. The second possibility seems less likely, for as he confided to Harriet, he believed that "Brooklyn is one of the worst places in the United States in which to bring up children, especially if they be intellectual." By then he and Harriet had two children, Alpheus Hyatt, born on June 28, 1901, and Katherine Goldsborough, born on September 9, 1903. Mayor opined that their children would be "surrounded by the stultifying average citizen, neither good nor bad . . . but self-satisfied, ordinary, and uninteresting." He told Harriet, "we must escape, and that very soon" from Brooklyn.[39] On the other hand, Mayor's aim in calling for the improvement of museums of natural history may have simply been part of his general desire to advance the quality of research in the United States, for shortly after he knew that he would be leaving Brooklyn, he published another article in *Science* that called for better salaries for professors, endowment of more professorships, and more funds for libraries and research.

By early June 1903 Mayor was in Nassau, where he collected until late July. Sponsored by the Brooklyn Institute, the expedition was quite successful. As Mayor reported in his "Medusae of the Bahamas," which he published in the institute's *Memoirs of Natural Science* in April 1904, he collected twenty-nine species of hydromedusae, six species of scyphomedusae, six species of siphonophores, and two species of ctenophores. He described two species of hydromedusae that he believed to be new. One of these he named *Parvanemus degeneratus* (=*Pachycordyle degenerata*), and the other he named *Lymnorea alexandri* (=*Podocoryna alexandri*), after his "constant friend" Alexander Agassiz. Mayor also prepared seven plates containing sixty-five black-and-white drawings, all done by his hand. As usual, his drawings were executed in fine detail, providing remarkable depictions of exquisite medusae. Mayor observed, however, that "the medusa-fauna of the Bahamas is poor in comparison with that of the Tortugas, Florida," the reason being, in his view, that the latter was close to the Gulf Stream, while the former was characterized by "vast areas of shallow flats, covered mainly with coralline mud . . . [and] a flocculent mass of silt."[40] No doubt, Mayor was partly correct in his assessment of the abundance of pelagic medusae in the Tortugas region, but he underestimated the potential of the Bahamas as a site. Despite his contention that he would objectively consider the Bahamas as a location for a marine station, he continued to favor the Tortugas site.

The anxious wait for a decision by the Carnegie trustees continued into October 1903, but near the end of the month Mayor received word that the trustees had invited him to discuss the matter with them. Perhaps Harriet had thought the proposal would not be approved, but now she saw that it might. She expressed concern that Alfred would be away from home quite often if he were appointed as the director of a laboratory at Tortugas. Mayor responded on October 26 by saying that her concern "has taken away the little desire I had for the position." It would, he admitted, entail a "hideous separation from March 1–August 15 every year." The position would, however, "give me the best possible chance" for ultimately receiving appointment as the director of the MCZ, which is "what we are really hoping for." If they viewed the Tortugas job as merely a step toward their goal of getting to Cambridge, then they could at least get away from "this hideous Brooklyn," he suggested.[41] It is not at all certain that Harriet shared Alfred's view of Brooklyn, for she seemed to be content keeping herself busy as a mother of two children, continuing her sculpting work, and enjoying involvement in community activities. However, of her devotion to Alfred there can be no doubt, and she would do whatever was necessary to see that he enjoyed his zoological work and that his career advanced.

The matter was settled in December 1903, when the Carnegie Institution of Washington formally established the Department of Marine Biology, to be built on Loggerhead Key. Alfred Mayor's dream of a biological laboratory based in the Dry Tortugas had become a reality, and he was to be its founding director.

# A Speck in the Sea

The exotic keys of the Dry Tortugas, where Mayor had met with collecting success and gained considerable inspiration, comprise an elliptical cluster of low, sandy islets and adjacent shoals and coral reefs in the southeastern Gulf of Mexico. Arising from the sea, the tiny archipelago is located some 70 miles (113 kilometers) west of Key West, Florida. The long axis of the formation extends over a distance of approximately 11 nautical miles (20 kilometers) and is oriented from northwest to southeast. Taken together, the islands cover little more than one hundred acres of dry land, although their areas and shapes vary due to the dynamics of erosion and accretion. The two largest are Loggerhead Key and Garden Key; others include Bush Key, Long Key, East Key, Hospital Key, and Middle Key, the latter two lacking vegetation. Bird Key existed at the time that the Tortugas Laboratory was established, but it later sank below the surface. A natural harbor is formed by the configuration of Garden Key and its surrounding reefs.[1]

Lying just across the Straits of Florida from western Cuba, the region derives its name from the Spanish "Las Tortugas," referring to the abundance of sea turtles found by Juan Ponce de Leon when he explored the area in 1513. Later the islands became known as the Dry Tortugas because there is no source of freshwater on any of them. They became American territory in 1821, when Spain ceded Florida to the United States.

Due to its geographic location, physiography, oceanography, and weather, the Dry Tortugas region has long been a major hazard to navigation in the Gulf of Mexico. In addition to sandy shoals, reefs, treacherous water currents, and occasional poor visibility, the islands have been badly battered on many occasions over the years by hurricanes and other major storms. Indeed, even though a lighthouse was erected on Garden Key in 1826, the area has been the site of more than two hundred shipwrecks. That original structure served as the primary

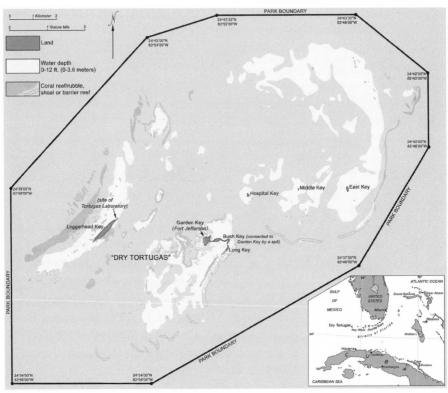

*Map of the Dry Tortugas, Florida, and Dry Tortugas National Park*

lighthouse of the area until 1858, when it was replaced by the 157-foot-high (48-meter) Dry Tortugas Light on Loggerhead Key. The lighthouse on Logger-head Key was there when Mayor established the Tortugas Laboratory and remains today as the principal lighthouse of the region. Its conical tower, constructed of brick laid on a stone foundation, is 8 feet and 9 inches (2.7 meters) thick and topped by an iron lantern. The tower, unpainted in early years, has been distinguishable for more than a century by its black upper half and white lower half. In 1876 a smaller lighthouse was constructed on Garden Key. Known as the Tortugas Harbor Light, it is 82 feet (25 meters) high and rests on the walls of a fortification known as Fort Jefferson. The Tortugas Harbor Light, now painted entirely black, has a hexagonal tower constructed of boilerplate iron. Both lighthouses were manned in Mayor's day but are now fully automated.[2]

Aware of the strategic location of the Tortugas islands at the entrance to the Gulf of Mexico, the United States began to build a major fortification on Garden Key in 1846. Named Fort Jefferson, it was designed to accommodate 450 cannons mounted in three tiers and to garrison fifteen hundred military and support personnel. Its brick walls, 8 feet (2.4 meters) thick, were in the form of an unequal hexagon: four of the sides were 476 feet (145 meters) long, and the others were 325 (99 meters) feet long. A saltwater moat was constructed outside the walls, and it was from that moat that Mayor and others would collect many species, especially the jellyfish *Cassiopea xamachana*.[3]

Fort Jefferson, the largest nineteenth-century American masonry fort, occupied a substantial part of the dry land on Garden Key. Construction proved to be difficult from the beginning, and progress was slow. Renewed military interest in the unfinished and already deteriorating fort began around the time of the American Civil War, and a second phase of construction was initiated, bringing the walls to a final height of 45 feet (13.7 meters). Because of its vulnerability to rifled cannons, however, the fort soon became obsolete and was never completed.[4]

Fort Jefferson remained in Union hands throughout the Civil War. During the conflict and for a short time afterward, it was used to house military prisoners. Among those confined to the fort's stockade was Dr. Samuel Mudd, who had treated the infamous assassin John Wilkes Booth. The U.S. Army pulled out of Fort Jefferson in 1874, and the structure had been reduced to a naval coaling station by the time of Mayor's first visits. Nevertheless, this military outpost and its personnel were of considerable value to scientists working in and around the isolated Tortugas.

The environmental fragility of this cluster of small keys has been recognized over the years. In 1908 President Theodore Roosevelt designated the Dry Tortugas as a wildlife refuge, in large part to protect the sooty tern, tens of thousands of which nest seasonally in the area. As Mayor and others noted, the Dry Tortugas are biologically important. The sea life is tropical in character, and its species composition marks these keys as part of what present marine biologists refer to as the Caribbean Biogeographic Province. Sedimentary bottoms, coral reefs, hard bottoms, and sea-grass meadows predominate among local marine environments. Swept by the warm Florida Current as it flows from the Gulf of Mexico into the Atlantic Ocean and representing a biologically productive oasis in shallow waters at the western extremity of the Florida Keys, it offers a wide diversity of plankton and more than three hundred species of fishes. The algal flora and invertebrate fauna of the sea bottom are also diverse. Especially conspicuous among the invertebrates are species of corals, sea fans, sea anemones, sponges, mollusks, crabs, sea stars, sea urchins, and spiny lobsters. As is apparent from the name of the islands, sea turtles of several species, including the loggerhead, green, hawksbill, Atlantic ridley, and leatherback, inhabit or visit the area and come ashore to lay their eggs on sandy beaches. These keys also provide an important habitat for some seabird species.

Naturalists had been interested in the fauna and flora of the Dry Tortugas decades before Mayor established a marine biological laboratory there. For example, John James Audubon visited these islets during the spring of 1832. The famous ornithologist described the Tortugas as comprising "five or six extremely low, uninhabitable banks, formed of shelly sand . . . [of interest mainly to] that class of men called wreckers and turtlers."[5] The so-called turtlers captured sea turtles and their eggs for food, while the "wreckers" were involved in the rescue or salvage of grounded or wrecked ships. By the 1850s the Harvard naturalists Louis Agassiz and his son Alexander had begun exploring the reefs and rich invertebrate assemblages of the area, and their work there continued off and on for several decades. Louis F. de Pourtalès, of the United States Coast Survey, visited the Tortugas during scientific investigations in the Straits of Florida between 1868 and 1878, and Charles Cleveland Nutting collected specimens around the Tortugas during the Bahamas expedition sponsored by the University of Iowa in 1893. A devoted naturalist, museum builder, and adventurer, Nutting was a strong supporter of Mayor's proposal to found a marine biological laboratory on Loggerhead Key and later named a species of hydroid in honor of Mayor (*Sertularia mayeri* Nutting, 1904).[6] Of course, Mayor had visited the Tortugas several times between 1897 and 1902, carrying out studies on medusae, undertaking experiments on the mating behavior of lepidopterans, and investigating the biology of the palolo worm.

The biological laboratory on Loggerhead Key became the first tropical marine research station in the Western Hemisphere. At the same time that the CIW established the Tortugas Laboratory in December 1903 it also approved a Department of Experimental Biology at Cold Spring Harbor, and Charles B. Davenport was to be its director. Mayor was pleased for his old friend, and he was delighted over his own good fortune. His campaign had finally paid off, thanks to his vigorous efforts and some help from Agassiz and Robert S. Woodward, CIW president, a physicist, an old friend of Mayor's father, and a firm believer that Mayor was the type of "exceptional man" mentioned in Andrew Carnegie's endowment. In a letter dated January 25, 1904, Mayor received the official offer, which provided for a salary of $4,000 per annum (equivalent to approximately $70,000 in the year 2000). The letter also specified an initial allocation of $20,000 for operating the Tortugas Laboratory during its first year, exclusive of the cost of a major seagoing vessel, and it requested a detailed plan for the buildings and equipment. Mayor quickly accepted the offer, and by January 28 he had drawn up a plan and an estimate of the costs. He requested an extension of time, however, saying that he could not begin on April 1 as he felt obligated to give the Brooklyn Institute trustees more time to find his replacement and as he must finish a book he had agreed to write for the New

York Zoological Society. The Brooklyn trustees held Mayor in great esteem, and they were unaware of his feelings toward their city. Striving to keep him, they offered to raise his salary immediately from the $2,650 he was paid in 1903 to $4,000 per year and to promote him to "Curator in Chief," which, as Mayor told Davenport, would place him in "absolute charge of everything." He added, "I really feel all torn up," for "they one and all came to me and *begged* me not to go." He would not change his mind, however, and he told another old friend, Herbert S. Jennings, that "they can't hold me back from Tortugas" and that he was "determined to do my best to make it a success."[7]

Indeed, Mayor would do his best, and he started with an ambitious plan. As he had wanted, the laboratory was to be set up on Loggerhead Key, near the north end of the little island, about 1,000 feet (305 meters) north of the U.S. lighthouse. The CIW received a license for the site from the U.S. Department of Commerce and Labor and the lighthouse board of directors. In his plan Mayor stated that the purpose of the laboratory was to encourage work "in embryology, physiology, histology, ecology, variation and other special directions of study in Zoology," which was to result in "extensive monographs . . . equal to the hithertofore unrivelled [*sic*] publications of the Naples Laboratory." For this purpose the laboratory would need a launch, a main laboratory building, a detached kitchen, living quarters, a storage shed, a cistern for catching rainwater, a windmill for pumping seawater and air to the laboratory, furnishings, and considerable equipment. It would also need a dock and a ship channel.[8]

Designed to accommodate as many as twelve researchers at any time, the laboratory building was to be a prefabricated wooden structure anchored by cables to protect against hurricanes. Mayor requested "a staunch, sea-worthy naphtha launch about 45–55 feet [13.7–16.8 meters long] on [the] waterline." It was to be fitted with auxiliary sails and a winch for dredging to a depth of 200 fathoms (366 meters). Mayor estimated the cost of the launch at $5,000, the buildings at $2,500, other items at $1,400, and annual operating expenses at $6,650, for an initial outlay of $15,550. His estimate of yearly salaries and operating costs did not include the expenses of investigators. The work of the laboratory would begin as soon as the research vessel had been constructed and as quickly as Mayor could pilot it south from Maine to Miami, collecting specimens on the way, particularly pelagic, or open-ocean, species. Then, when the laboratory was ready, he would begin "a thorough collection of the animals of the reefs and shallow waters," launch an inquiry into "the growth of corals under various conditions," trawl daily for pelagic fauna, and conduct some specific studies he had begun. In his view, he should, as director, devote about six months of each year to the laboratory and spend the remaining months at other research stations. After four years on that schedule, the director would give all of his time to the laboratory. No doubt, his grand plan impressed the Carnegie trustees, but it would have displeased Harriet, although it is not certain that Mayor told her all of the details of his proposal. He did hope, however, that facilities on Logger-head would eventually accommodate spouses and children—a hope that proved to be as unrealistic as the idea of eventually keeping the laboratory open year-round.[9]

Although Mayor grossly underestimated some of the difficulties of placing a laboratory on a relatively tiny and remote island, the trustees approved his plan, and by late March a 57-foot (17-meter) auxiliary ketch, with a 3.5-foot (1-meter) draft and powered by a twenty-horsepower naptha engine, had been built. Capable of dredging to 500 fathoms (914 meters), the vessel was designed by Stearns and McKay of the Marblehead Yacht Yard in Marblehead, Massa-chusetts, and constructed by the Rice Brothers Company in East Boothbay, Maine. It was larger than Mayor had originally requested and cost an additional one thousand dollars. The charge for the prefabricated and portable laboratory, constructed in New York by the Drecker Company, also ran slightly higher than Mayor's estimate and did not include the expense of shipping the sections to Loggerhead, which ran to four hundred dollars. Working feverishly, Mayor had made arrangements by June 22 to send the sections to Tortugas around the first of July. Meanwhile he went to the island with a crew to clear and level the site. While the area was being cleared, a fire spread from a brush pile and threatened to denude the area of its scattered vegetation, but Mayor saved the day by order-ing the men to make a firebreak. He was less fortunate in receiving the prefab-ricated sections of the building, however. Contrary to his instructions that all sections were to be sent on a single steamer, the shipper divided them into two lots, the second of which did not arrive until two weeks after the first.[10]

By the end of July all the buildings were in place, and by late August an 84-foot (25.6-meter) dock was complete. The main building of the Tortugas Laboratory was L-shaped, with the north wing designed as a laboratory and the east wing for private quarters. A double roof, with air space between the sec-tions, and a ventilator in the ceiling of each room were intended to cool the facility. Mayor designed the roofs to be loosely attached so that hurricane winds would blow them off without placing great strain on the building walls. He arranged to fix the main building by cables to T-shaped anchors, which, he hoped, would make "it well-nigh impossible to overturn the structures" if struck by high winds. Resting on wooden posts driven deep into the sandy soil, the floor of the building was 2 feet (0.6 meter) above the ground. Because wind vibrations would affect work done by microscopes, Mayor had holes cut through the floor for the legs of eight tables that would hold those instruments. Other facilities included a dining room and library; a stand-alone kitchen; a small shed for alcohol, naphtha, and chemicals; and a galvanized metal cistern for catching rainwater. The buildings were amply furnished, and the laboratory contained microscopes, a balance, dissecting instruments, and a large supply of glassware and other necessities. Believing that "hygienic precautions are absolutely necessary if northern men, unaccustomed to a hot, moist climate, are to maintain good health," Mayor noted that "modern plumbing" was introduced.[11]

At last Mayor was ready to get the laboratory in operation, and he planned to open it in March 1905. He would personally pilot the yacht from New York to Tortugas. When the vessel was launched in late August 1904, Mayor dubbed it the *Physalia,* after *Physalia physalis,* a siphonophore commonly called the Portuguese man-of-war. The species is floating, wind-driven, and abundant in tropical waters, including the Gulf Stream, and is well known for its venomous sting. Mayer had painted the species in 1893, early in his investigations on hydrozoans. *Physalia* was launched on August 19 at a cost of $6,037. Mayor had the beautiful yacht equipped with reagents, glassware, dredges, trawls, nets, and other equipment. A 15-foot (4.6-meter) tender powered by a naphtha engine was carried on board. With a small crew and Mayor in command, *Physalia* sailed from East Boothbay on August 24 and arrived in New York on September 25, with a week having been taken for fitting out in Gloucester, Massachusetts. Mayor also collected specimens along the way.[12]

As autumn of 1904 set in, Mayor was preoccupied with the task of taking *Physalia* south. He departed New York on October 15 and arrived in Miami, Florida, on January 23, 1905. Mayor had made part of the journey through intracoastal waters, and he collected specimens as opportunities arose. He also prepared color drawings of living jellyfishes and comb jellies observed during the cruise. Even as he was making preparations to establish the laboratory at Tortugas, Mayor was considering the possibility of resuming work on his book on medusae, hoping that he could get it published under the auspices of the CIW. First, however, he must find out what Agassiz intended to do. In response to his query about the matter, Agassiz wrote to him on August 6, 1904, saying, "you can do with it [the manuscript] what you please" and that he "must give up all connection with it" because he had no time to work on it. He told Mayor that he could mention his name in the preface but could not list him as co-author. A few days later Agassiz told Mayor, "you don't seem to understand that while I am ready to have you publish . . . [the book], I did not and cannot agree to your stating that I agreed to your publishing it in the Carnegie Publications." Agassiz maintained that because he had associated himself with the CIW, he might be viewed as seeking a favor if his name appeared as coauthor, but he did not seem to recognize that Mayor's association with the CIW was even stronger than his own. The streak of stubbornness ran deep in the character of Agassiz, however, and he could be downright ornery at times. In any case, Mayor was free to proceed on his own to finish the book, a work that in reality was almost solely due to his own efforts.[13] Much remained to be done, however, before it was ready for publication, and Mayor would have to devote most of his attention to the Tortugas Laboratory during the next several months. The months turned into years before the monumental task on medusae was done. In the meantime, however, Alfred Mayor would be working hard to meet his goal of developing the Tortugas Laboratory into one of the best marine research stations in the world.

Much work remained to be done before the first researchers were to arrive in 1905, however, including publication of the popular book Mayor had hastily written for the New York Zoological Society. In late September he found that the society "had done *nothing*" to publish his manuscript. Perturbed, he told Harriet that he intended to withdraw the manuscript and try to place it with the Macmillan Company for a single royalty of $250, though the society considered the value of the work to be at least $1,000. His displeasure apparently prompted Zoological Society officials to get busy, for the little book appeared in print in 1905 as a volume in the New York Aquarium Nature Series. Intended for "the beginner," *Sea-shore Life: The Invertebrates of the New York Coast* aimed "to present in clear, untechnical language, a description . . . of the larger and more conspicuous marine invertebrates of the coast of New York State." It included 119 figures, of which 111 were photographs, all but one taken by Mayor. Mayor introduced the volume with a statement in support of the theory of evolution, and he endorsed it again later in the book. After referring to the "constant competition for life" among invertebrates and noting that "the weak and unfit must perish," Mayor stated that new species may appear "suddenly . . . [and], through inheritance, perpetuate their new peculiarities." Anyone interested in zoology, he added, "should read Darwin's 'Origin of Species' . . . [for] this work is to the natural sciences what Newton's 'Principia' is to the physical and mathematical sciences."[14]

Unfortunately, *Sea-shore Life* contained errors, reflecting, as a reviewer said of it, "signs of hasty preparation," especially in the sections on crustaceans and mollusks. He added, the book "shows a carelessness in arrangement and constant misstatement of facts." In a letter to his wife Mayor admitted that the book "ought to be better." The criticisms were valid, of course, but they unfairly treated the volume as though it were a scholarly monograph instead of an elementary, popular book. Many years later Charles Townsend, Mayor's old friend and director of the New York Aquarium, noted that *Sea-shore Life* contains "certain typographical and other errors," but he observed that these were due to its being "hurried into print" while Mayor was away. Despite its shortcomings,

*Sea-shore Life* served a useful purpose, and it sold so well that it was reprinted without revision in 1906, 1911, and 1916.[15] Mayor likely realized no profit from the book, however, for it appears that he gave the rights to it to the Zoological Society.

By mid-December of 1904 Mayor had compiled a list of noted scientists he wished to invite to the new laboratory at Tortugas for research between April 1 and August 1, 1905. In order to attract these accomplished biologists to the remote island, he told CIW secretary Charles D. Walcott that the Carnegie Institution must offer to pay their expenses and promise to publish the results of their research. Having received no response from Walcott by early February, he turned directly to CIW president Robert S. Woodward, saying, "I regard this matter as of *great* importance." Approval soon followed, and Mayor issued the invitations. When the acceptances were all in, Mayor had seven biologists lined up: William K. Brooks, Edwin G. Conklin, H. S. Conard, Rheinart P. Cowles, Herbert S. Jennings, H. F. Perkins, and Jacob E. Reighard. In addition the scientific collector Davenport Hooker and Brooks's assistant Carl Kellner would come along. All of the researchers were zoologists except Conard, who was completing a doctorate in botany at the University of Pennsylvania.[16]

Each of the researchers arrived and departed individually between April 21 and July 26, thus requiring Mayor to take the *Physalia* to Key West or Miami many times. In a letter to Harriet on May 2 he said, "I am savagely busy. . . . The worries of this job are far greater than those of Brooklyn," but he assured her that he would "get them all straightened out in time." His optimism was hard to maintain, however, for problems occurred from the outset and continued through much of the first year of operation. Among them was an encounter with a hurricane in mid-November 1904, when Mayor was sailing the *Physalia* from New York. The ship lost its jib and had to be towed into Charleston for repair. Later, in Miami, Mayor discovered that iron on the rudder and centerboard had corroded, and the *Physalia* had to be hauled into dry dock for repairs. In late May 1905 another storm lashed the laboratory and blew off a ventilator —a harbinger of worse to come as recurring natural phenomena played havoc with the man-made objects placed on Loggerhead Key.[17]

In late June 1905, shortly after he had arrived at the laboratory, Conard became extremely ill from dysentery, and Mayor decided that he had better take him to Key West. Shortly after setting sail, the *Physalia* broke down, forcing Mayor to return to Loggerhead Key and go to Fort Jefferson to telegraph a request for the navy to send a tugboat from Key West. Mayor accompanied the desperately ill Conard on the tug, but about 8 miles (1.3 kilometers) from Key West the boat "ran aground on a coral reef . . . in a pitch black night." For reasons unknown, a few women were aboard the tugboat, and, in Mayor's words, they "began to pray and jump around as if they were scared cats." He calmed them in due course, assuring them that the tug could not sink since it was already "hard aground." When the tide rose, the tug freed itself, and it reached Key West just in time for Mayor to get Conard aboard a steamer heading for Tampa. In Tampa, Mayor placed the ailing man on a train, paying his fare and tipping a porter to look after him. Mayor told Harriet that this was the sixteenth problem he had encountered since leaving home for Tortugas—and he listed each of them. There would be others before he got through the season.[18]

In his second annual report to the trustees of the CIW, Mayor noted some of the problems and said, "the difficulties of maintaining an efficient laboratory at Tortugas are considerable," but he hastened to assure them that the difficulties were "by no means insurmountable." Once again he lauded the Tortugas station by noting that "the pelagic fauna is the richest in the Atlantic." He noted that he had beautified the grounds by planting one hundred coconut palms and some date palms, azaleas, rubber trees, banana trees, and cacti. This report reveals how much was actually accomplished in establishing the laboratory that year. A 22-foot (6.7-meter) Swampscott dory, with a four-horsepower motor, was put into service for daily surface tows and dredge samples. Moreover, Mayor managed significant construction projects during the year. Two buildings were built, one on either side of the dock. The first of these was a kitchen, and the second was an aquarium-laboratory that was outfitted with tanks and a water pump. Two additional cisterns, holding 1,326 cubic feet (37.5 cubic meters) of freshwater, were completed, and a windmill to power delivery of saltwater to a saltwater tank and to bathrooms was set up. A below-ground oil house, to store fuels, alcohol, and other flammable materials, was built. As Mayor indicated in the report, the laboratory "now consists of two main portable buildings, the kitchen, aquarium, dock, oil-house, out-house, wind-mill, and three cisterns, while the vessels are the 60-foot [18-meter] 20-horsepower ketch *Physalia,* the 22-foot [6.7-meter] 4-horsepower launch, the 15-foot [4.6-meter] ¾-horse-power launch, and a row boat."[19]

Meanwhile, Mayor had initiated a study of a subject that continued to fascinate him for the next decade. Having observed the pulsating movements of small jellyfishes, or hydromedusae, and especially the larger, true jellyfishes, or scyphomedusae, he came to believe that their rhythmic motions resembled the pulsations of the human heart. Thus, by May 1905 he had begun a series of experiments designed to reveal information on the physiology of medusae. In early June 1905 Mayor told President Woodward, "I have long waited for the opportunity to go into physiological work, wherein the greatest advances in zoology are . . . to be made in the next few years." He added, "this station affords a unique advantage for such work."[20] Mayor was indeed excited over his research on medusan pulsation, and he sought to explain some of his initial discoveries to Harriet in four of the letters he wrote from Loggerhead Key during the summer of 1905. In a series of experiments he severed the nerves that ring the fringe of a medusa. He was able to make the medusa pulsate after touching it with a crystal of potassium sulfate. "I think," he informed Harriet, "we can make the heart 'go' in very much the same way," and he predicted that surgeons might someday "take a man's heart out and put it back after operating on it." If that were so, then "many forms of heart disease would lose their terrors."[21]

By 1906 Mayor had completed his study of "Rhythmical Pulsation in Scyphomedusae" and published it as a CIW monograph. Particularly noteworthy in this report was an account of his discovery of what behavioral physiologists refer to as the "entrapped wave" preparation (see page 117). The American physiologist L. M. Passano has reviewed details of this elaborate preparation, used for experimental studies on conduction of nervous impulses.[22] Although he experimented with the hydromedusa *Gonionemus* and with the scyphomedusae *Aurelia, Dactylometra* (=*Chrysaora*) and *Cassiopeia* (=*Cassiopea*), Mayor conducted most of his studies on the last of these, a genus of scyphomedusae commonly referred to as "upside-down jellyfishes." This descriptive name is entirely appropriate because of the habit of these medusae of lying upside down in shallow tropical waters, thereby allowing light to reach symbiotic algae in the tissues of their undersides. In addition Mayor conducted preliminary experiments to determine the effects of depriving *Cassiopea* of nutrients and sunlight, the latter of which was intended to determine the role of the zooxanthellae, or algae, so prominently observable in *Cassiopea* but also present in certain other cnidarians (especially reef-building corals).[23]

In the meantime, apparently in 1905, Mayor had teamed up with Caroline G. Soule to conduct a series of experiments on the behavior of certain caterpillars and moths. Just how they got together on the subject is unknown, but they obviously shared an interest in butterflies and moths. Their study was published in 1906 in the *Journal of Experimental Zoology.*[24] This was the last research on Lepidoptera published by Mayor.

Within two years after establishing the Tortugas Laboratory, Mayor had come to realize that one drawback to its location was the lack of accommodations for the families of the researchers. He may have recalled a critic's comment two years earlier that the Dry Tortugas "seems to be an ideal place for solitary confinement." As early as March 24, 1905, Mayor told Davenport that he doubted the practicality of accommodating wives. "The hot, moist climate is probably too uncomfortable for most women," he said, "and their presence would destroy the freedom and comfort of the men." Eventually, however, women might be accommodated during the winter season, he opined. As he

began to line up researchers for the second season at Tortugas, Mayor ran head-
long into the problem of separating researchers from their families when
Thomas H. Montgomery asked: "Could a whole family live there [on Loggerhead Key] in tolerable comfort?" More pointedly, he queried: "Do you have to
forsake Mrs. Mayer during the summer, or does she accompany you?" Of
course, since some of the investigators came for only two or three weeks, the
length of separation for them was bearable, but for those who spent up to six
weeks or longer there, the time away from family was stressful. Obviously, no
one felt the strain more than Mayor, for he usually left for Tortugas in March
and did not get back to his home until late August or sometimes early September. Indeed, he was at the marine station when he received a telegram in late
May 1906 notifying him of the birth of his third child, named Brantz.[25]

Separation from family was but another problem of having the marine laboratory located in such an isolated place, and in a letter dated March 23, 1906,
Mayor told Harriet that he intended to suggest to Woodward that it be moved
to Nassau, where costs would be lower and investigators "could bring their wives
and families." He did, in fact, send the suggestion to Woodward, but although
Woodward's response is unknown, the CIW president likely told Mayor that the
trustees might be unreceptive, given the case that Mayor had made for Tortugas.
In the meantime, though unknown to Woodward or the trustees, Mayor was
still holding out hope for becoming director of the MCZ. As the scholar Mary
P. Winsor has observed, "Billy" Woodworth had failed "to command the respect
of others" and had been demoted in 1904. Leadership in the museum remained
in flux, and Mayor believed that he was the man to straighten things out. He
thought his chances were slim, however, for he believed that Agassiz still controlled the MCZ and was not in favor of bringing him back.[26]

A month later Mayor was certain that he would never get the position, noting in a letter to Harriet on April 28, 1906, that his former instructor Edward
L. Mark "is insanely jealous and hates me secretly." He also lamented that his
most "powerful friend" at Harvard, Nathaniel Southgate Shaler, his former geology instructor, had died recently. It is not clear why Mayor now considered
Mark to be his enemy. Perhaps he had come to view his former professor as a
rival since Mark had established the Bermuda Biological Station in 1903. As further proof of his belief that he had no chance for the MCZ position, Mayor told
Harriet he was sure that Agassiz had persuaded two MCZ associates, Henry B.
Bigelow and George H. Parker, to decline Mayor's invitation for them to conduct research at the Tortugas Laboratory after each had "practically accepted."[27]

Mayor nevertheless looked forward to opening the laboratory in 1906. Before going on to Tortugas, however, he spent a few days in the Everglades, using
a canoe to negotiate around the thickets of mangroves to collect specimens. He

was particularly delighted to collect specimens of *Velella velella,* a lovely little floating blue porpitid hydroid, commonly known as the "by-the-wind sailor." An inhabitant of open ocean waters, these individuals had somehow drifted into the mangroves and provided a lucky find for Mayor. On March 28 Mayor reached Loggerhead Key and "found everything in 'first class' order." Only a day later, however, while the crews of the *Physalia* and the smaller vessel *Sea Horse* were playing out the mooring chains, the "beautiful clear sky" suddenly gave way to "a black cloud . . . [and] a dull, heavy roar." In a few moments the men faced "a storm of almost hurricane violence." A launch towed by the *Physalia* broke loose, washed onto the shore, and was soon "dashed into kindling wood." Swirling around the vessels were "great black water spouts," and Mayor and his men watched as the fierce wind ripped roofing from the laboratory, lifted barrels and boxes into the air, and broke off bushes. The storm raged for six hours, but the men escaped injury. By the middle of June four other storms had struck Loggerhead Key, but none was as severe as the first.[28] Clearly, however, the laboratory was at the mercy of forces beyond the control of anyone.

As he looked back on 1906, the second year of operations of the Tortugas Laboratory, Mayor could suppress memories of the problems he had encountered and reflect on his successes, though the latter had not been as great as he had hoped. In his own case, however, he was pleased with the progress he had made on the manuscript of his book "Medusae of the World." He had completed "100 drawings of new and rare medusae." Moreover, he had collected many specimens of siphonophores from the Gulf Stream. By November 1906 he was working steadily on the book, but as he told Davenport, the project "seems to become larger an undertaking every day and less and less worth while," adding that "long it certainly will be." As a dedicated systematist, Mayor believed that he needed more specimens and more data in order to develop the work into the classic he had once promised to make it. Thus, he decided to do more collecting from the deep waters around the Bahama Islands during most of April and half of May 1907 and to study specimens at the Stazione Zoologica in Naples from November 15, 1907, to January 15, 1908. His journey to the Italian laboratory would be part, as the historian of science Jane Maienschein said, of the "long tradition of American expeditions to the Naples Station." Indeed, Mayor referred several times to the Stazione Zoologica as the model for American marine laboratories to emulate or, in his view, to "surpass."[29]

On March 30, 1907, Mayor set out in *Physalia* for Cay Verde, some distance from Nassau. The main purpose of the trip was to take Frank Chapman to the area to study birds. Along with him was George Shiras III, a pioneer in wildlife photography. Shiras later wrote an account of a fearful adventure that soon followed. That story corroborates and supplements the awesome account related by Mayor in his annual report to the CIW and in a letter to his wife. The weather was calm early during the day of April 1, but when the *Physalia* was about fifty miles (eighty kilometers) southeast of Nassau, a wind began to blow from the south and soon turned into a powerful gale. Around 4:00 P.M., unable to make progress against the wind and surging waves, Mayor piloted the yacht "into a small harbor between two small islands called the Elbus" and waited for the storm to abate. An hour later, however, "a huge heavy cloud blackened up in the north," and within moments a major hurricane-force storm was threatening to smash the *Physalia* on the reefs. Mayor quickly ordered anchors lifted, but almost instantly the vessel became stranded on a bar. A huge wave smashed across the deck and flooded the engine room. Then, in rapid succession, two large breakers suddenly lifted the yacht high and pushed it over the bar into deep water. Mayor managed to get the engine running again and guided the ship into the "inky black" night at seven knots an hour.[30]

The "blind run," as Mayor called it, was especially precarious because of the unseen coral reefs. Near midnight a fuel tank burst, and the escaping naphtha flooded the hold. After extinguishing all lights for fear of igniting the volatile fuel, Mayor steered the ship by use of a small "pocket lamp." As dawn broke and the storm passed, he discovered that the vessel had been blown some 80 to 90 miles (130–145 kilometers) from the place where it had sought refuge. Finding a small island, Upper Golden Ring Key, Mayor anchored *Physalia* and had the hold pumped out. Two days later he headed the yacht for Nassau over rough seas, aiming to get two members of his research team to Nassau in time to catch a steamer. Because only a few gallons of fuel remained in another tank, Mayor had to hoist sail and tack against the wind. He also decided to sail at night, but on April 8, within 3 miles (5 kilometers) of its destination, *Physalia* "ran hard upon a sand bank and pounded in an angry sea all night." There the vessel remained for three days, being freed only after it was "stripped of her ballast." At last *Physalia* sailed into the harbor of Nassau, and, said Mayor modestly, "thus ended as adventurous a cruise as I have ever had."[31]

After several days of collecting medusae from the waters around Nassau, Mayor went on to Loggerhead Key, arriving there near the end of May. Lack of rainfall since the previous November and "the merciless sun" had resulted in the death of many plants on the island, and the water supply in the cistern had fallen greatly. Undeterred, Mayor moved ahead. A dozen researchers came and went during the season. Conklin and Reighard were among those who returned. Frank Chapman was also there again, and Mayor arranged to transport him about the region to study the nesting habits of the magnificent frigate bird, *Fregata magnificens* (=*F. aquila*), and the northern gannet, *Sula bassanus*. Also present was John B. Watson, a professor at Johns Hopkins University. Mayor was particularly impressed with Watson and his behavioral studies of gulls and terns on Bird Key, one of the smallest islands of the Tortugas group. Mayor praised Watson for working "under circumstances so trying that but few men of science would have the resolution . . . to conduct the experiments." Watson had stayed on Bird Key for two months when the temperature of the sand soared to 120°F (49°C), finding occasional refuge from the searing sand and spending nights in an abandoned building that had been the yellow-fever hospital when Bird Key served as a quarantine station. He was not entirely isolated, however, for he kept four monkeys with him, allowing them the run of the islet and providing him, or so he believed, an opportunity to study the behavior of uncaged simians.[32]

Another scientist working at the Tortugas station in 1907 was Davenport Hooker, of Yale University. He served as collector for the station but also conducted studies of newly hatched loggerhead turtles. Robert Hartmeyer, of the Berlin Zoological Museum, was also there, investigating ascidians (sea squirts) and echinoderms (sea urchins, sea stars, and relatives). In addition Charles Stockard, a professor in the Cornell University Medical College, studied tissue growth and insects, and Mayor devoted time to further studies of *Cassiopea* and the palolo worm. He also worked on the medusae manuscript, telling Harriet in a letter on June 28, "I am making great progress with the Medusa book." In fact, he had completed the first volume, but he had a long way to go yet. The industrious zoologist also kept himself busy by preparing lectures to present at the Cold Spring Harbor station and at the MBL in mid-August. In addition he worked up four papers to present before the International Zoological Congress, scheduled to meet in Boston on August 19–24, 1907. He informed Harriet that he would do the latter "so as to advance the reputation of the laboratory."[33]

Mayor was pleased with the presentation of his papers at the meeting of the congress, but he came away somewhat unhappy over the behavior of Alexander Agassiz, who was serving as president of that organization. In Mayor's view, Agassiz "made such an ungraceful presiding officer that everybody was disgusted with him." Noting that Agassiz had come to the congress directly from a disruptive meeting of his copper-mining company and was thus "mad as a lobster," Mayor related to Harriet that Agassiz vehemently opposed a motion calling for the congress to study the impact of the Panama Canal on the area's fauna. Agassiz, he reported to Harriet, called the motion "perfectly ridiculous" and sat down, his face flushing red in anger. The old man had also refused to recognize two distinguished zoologists who rose to speak. As Mayor viewed it, Agassiz's comments represented "a purile [*sic*] attack on modern zoology." Later, at a dinner for Harvard faculty and alumni, when asked to offer comments on his work, Mayor boldly stated that Harvard was in danger of "sink[ing] to a lower level in zoology."[34] His temerity was hardly likely to enhance his chances of eventually becoming director of the MCZ, but it was the mark of a man who was more and more willing to speak his mind.

In early October 1907 Alfred Mayor embarked for Europe. Harriet and the children and Harriet's sister Anna Hyatt were already there, having arrived in June. They would remain there for a year, spending much of their time in France but also traveling to Britain and Italy. Harriet, still suffering from a persisting cough, came mainly to improve her health. The stay also provided an opportunity to enhance the children's education and to be with Mayor while he worked at the Stazione Zoologica. Anna Hyatt, who was rapidly gaining notice for her animal sculptures, wanted to study with famed sculptors in France.[35]

Before going on to Italy, Mayor decided to collect medusa specimens in the waters around Cornwall, England, and Harriet, pregnant with their fourth child, came with the children to join him in Cornwall. By late November 1907 they were in Naples, where Alfred stayed until the end of February 1908, spending his time doing further research on jellyfishes and working on the manuscript on medusae, completing sixty-seven figures during his stay. By the end of the year in 1908 he had finished the work. The manuscript, he noted, was "1387 pages and 700 + figures long." Seeing the work through the press would take a considerable amount of time, however, and two years would pass before the grand work appeared in print. During the intervening time Harriet had miscarried, but nothing else is known about the unfortunate event.[36]

At some point during this busy period, Mayor managed to write a popular piece for the *National Geographic Magazine,* which published it in 1908. It is likely that Mayor drafted the piece for three reasons: (1) to publicize the Tortugas Laboratory; (2) to satisfy his penchant for picturing the natural world; and (3) to express a genuine concern over environmental changes in Florida and along the eastern coast of the United States. An expert with the camera, he included several photographs with the article. Titled "Our Neglected Southern Coast: A Cruise of the Carnegie Institution Yacht 'Physalia,'" the article was based on Mayor's recollections of scenes encountered during the cruise of the vessel from Maine to Miami in 1904. He wrote lovingly of "the tall trunks" of cypress trees standing "like temple columns" and the "gnarled live-oaks, with their funereal pendants of Spanish moss dangling like old gray beards from the boughs," but he warned of "the impending ruin of the forests" of Florida. In addition Mayor lamented "the wanton destruction of interesting animal life" in that state. Although he devoted most of his attention to Florida, Mayor also sought more generally to depict the "mysterious attraction" of the Atlantic coast from Norfolk, Virginia, to the tip of Florida, and he expressed a strong desire to preserve its beauty.[37]

By mid-April 1908 Mayor was back at the Tortugas Laboratory ready to welcome the first of nine scientists who would conduct research at the station during that season. Some of them would be there for the first time, including the University of Chicago biologist W. L. Tower, the Columbia University bryozoologist Raymond C. Osburn, and the United States Geological Survey (USGS) scientist Thomas Wayland Vaughan. Tower brought seeds, plants, and barrels of earth with him in order to study the development of introduced flora. He was also interested in the evolution of new traits in insects, particularly in a genus of beetles. Already an acknowledged authority on corals, Vaughan commenced a series of studies of the development and distribution of reefs, including not only those around the Florida Keys but also those near the Bahamas. Mayor was strongly impressed by the work of Vaughan, who would return regularly to the Tortugas station to continue his research, though not without creating stress by his fussy and fastidious manners and his intense nature.[38]

In his report to the CIW trustees for 1908, Mayor again trumpeted the value of the Tortugas Laboratory, saying that it offered "unrivaled facilities for the prosecution of researches such as have not hitherto been conducted within the tropics." He also stressed the necessity for doing studies of the physiology, embryology, behavior, and evolution of marine animals. The major aim of the laboratory, he said, was to investigate "the laws governing life, rather than . . . the systematic collection of groups which have already been extensively studied." In any case, although he mentioned studying "laws of heredity" in the mission of the laboratory, Mayor did not cite others. Given his clear interest in the theory of evolution and his stress on embryology and physiology, he probably hoped that Tortugas researchers would at least frame some of their conclusions in such a way that generalizations about biological developments could be learned, but he did not elaborate. To attain the goal of the Tortugas station, Mayor said that researchers were selected "from among the most promising of our country's productive students."[39]

Although he mentioned the increasing costs of running a remote laboratory, Mayor said nothing else about moving it to another place, though he noted that the facility was located "upon the most inaccessible island along the entire range of the Atlantic coast." He also observed that the "isolation" of the laboratory ran up the costs of operating it, but he added that its seclusion offered researchers "absolute freedom from interruption."[40] Indeed, he was right on both counts: the costs were quite high, and the scientists who spent time at the laboratory were in no danger of facing human interruption. In all likelihood, however, the isolated men would have welcomed an occasional break from the solitude. Mayor's generally optimistic report could hardly be read by the CIW trustees, however, as other than a clear indication of the prosperity of the laboratory.

Mayor had not abandoned the idea of moving the laboratory, nor had he given up the thought of finding another position. In fact, soon after he returned from Europe, he told Harriet that he had spoken with Henry F. Osborn, president of the American Museum of Natural History, who informed him that a position at the AMNH might open up for him. "Certainly, my destiny," he wrote to Harriet, "is toward New York or Washington, for in both places I am gaining ground." He was, however, losing ground at Harvard, and he would not be helping his chances there, he added, for he intended to accept an offer to serve as editor of the section on marine biology for the *American Naturalist,* a major journal. Since the editors of the *American Naturalist* had not sought the advice of Agassiz in selecting him, Mayor believed that "this choice will put 'Alex' in a new rage and still further decrease all [my] chances at Harvard."[41]

A few weeks later, however, Mayor reported a development that could affect Agassiz's continuing control of the MCZ. Recipients of doctoral degrees from Harvard had recently been recognized, said Mayor, as "real alumni" by being granted the privilege of voting for the university's overseers, or trustees, a measure that Mayor viewed as previously defeated by Agassiz. "Alex is getting hit hard," he said, as he gleefully announced to Harriet that he had voted for two candidates who, presumably, resented Agassiz's continuing influence. "Whew, how Alex would swear," added Mayor. He also noted that Vaughan had dismissed "Alex's paper on the coral reefs . . . [as] not worth the powder required to blow it to H—!"[42] It is uncertain that Vaughan used the expression attributed to him by Mayor, but it is a fact that he severely criticized Agassiz's hypothesis— and for good reasons. Mayor continued to hold conflicting views of the man who had helped to bring out and develop his own ability but who had also exploited him and stood in his way, or so Mayor believed. Perhaps Mayor was exaggerating his notion of Agassiz's influence and his belief that the old zoologist was standing between him and a position at Harvard, but it is likely that Agassiz did not like the independence Mayor had finally but firmly achieved.

From time to time Mayor published comments on matters related to the improvement of biological research. In a 1906 issue of *Science,* for example, he criticized universities for promoting summer schools, thereby pressuring already low-paid professors to teach year-round rather than having a period for research. American universities should be following the lead of their German counterparts by fostering research, he declared. In 1908 Mayor published "Marine Laboratories, and Our Atlantic Coast" in the *American Naturalist.* To Harriet he had confided that his proposition "throws a bombshell into the Whitman camp." Charles O. Whitman had been arguing for the concentration of research monies and energies in the MBL, and Mayor was calling for a marine station in each of the four "faunistic regions" of the American Atlantic coast, of which the southernmost, from Biscayne Bay to the Keys, had already been established at Tortugas. A few months later Mayor returned to the theme of university responsibility for promoting research, arguing, once again in the pages of *Science,* that the aim of universities must be "to discover," not merely "to teach." He repeated his belief that too many young faculty earned so little that they were compelled to teach during the summers, when they should be conducting research. Perhaps Mayor wrote these articles because he genuinely desired to advance research in America, but it is also likely that he hoped they would help him ultimately to win a post in a major museum or university, as president of the Carnegie Institution or "some other tip top job." To help himself in reaching his goal, Mayor informed his wife, he must become a member of the Cosmos Club, a group of distinguished scientists in the nation's capital. As "a climber," he added, he could gain greater recognition by being counted among the club's members. Wayland Vaughan promised to nominate him for "associate membership" for the period he was to reside in Washington, that is, during the winter of 1908–9.[43]

Prior to opening the laboratory at Loggerhead Key in the spring of 1909, Mayor arranged for ten tons of New Jersey soil to be hauled to the island and spread about the laboratory grounds, the purpose of which was to break the eye-straining glare of the sand and to provide fertile soil for introduced shrubs and trees that would make the place seem less forbidding. Among the specimens he brought on this occasion were hibiscus, bougainvillea, poinsettias, and coleus. "We hope," he optimistically reported, "that in a few years the buildings may be embowered in a beautiful tropical garden." Meanwhile he was growing more concerned about the increasing isolation of the laboratory, for the U.S. Navy was beginning to abandon its base at nearby Fort Jefferson, on Garden Key. In a letter to President Woodward on April 21, 1909, Mayor noted that the cable, the wireless, and the water tanks had been removed from the base and that the navy was now running vessels there "very infrequently," thus requiring him to sail the *Physalia* more often to Key West for supplies and to transport researchers. Mayor sought to assure Woodward that these changes would have no appreciable effect, but two days after that he told Harriet that "the Tortugas is now actually more isolated than Australia." Only two weeks later he informed Woodward that "we must have a swift ocean-going vessel here, costing about $15,000, to make the trip to Key West every week throughout the season or we cannot keep the laboratory here." Once again Mayor was thinking of asking Woodward and the CIW trustees to move the laboratory to a more accessible location, but he said little about the matter until he filed his annual report at the end of the year. In the report he indicated that, while collecting jellyfishes around Jamaica in March, he had concluded that "the varied fauna and flora of Jamaica set it apart from Tortugas in its superiority." It was not a good time to push for moving the laboratory and thereby requesting additional funds, however, for the CIW president noted that "the expanding needs" of the institution's ten departments "must soon press closely upon [its] income." Following "the strictest economy" would be necessary, he added.[44]

Mayor had more to worry about than the problems of operating the laboratory during the season of 1909, for Harriet's cough had worsened. Almost certainly both Mayor and his wife now feared more than ever that she had tuberculosis, but neither wanted to admit the probability. "Those lungs probably have not consumption," Mayor wrote to Harriet with faltering assurance, "but some sort of influenza, or possibly even the liver fluke."[45] A woman of resilience and patience, Harriet encouraged Alfred not to worry but to give his attention to the operation of the laboratory and to his own research. The task was difficult, but Mayor proceeded, while maintaining hope that Harriet's condition would take a favorable turn.

Among the researchers who worked at the Tortugas station in 1909 were Charles Stockard, W. L. Tower, T. Wayland Vaughan, and five newcomers, including E. Newton Harvey, a University of Pennsylvania professor who would return several times thereafter to conduct studies of bioluminescence in marine invertebrates. During the year Mayor managed to bring out the first two volumes of *Papers from the Tortugas Laboratory*. The first volume contained ten papers, and the second, nine. Other papers resulting from studies conducted at the laboratory through 1908 were published elsewhere. Mayor reported that a total of thirty articles had already been published by twenty-nine investigators working at Tortugas, not counting twelve of his own.[46] Such productivity indicated great success for the Tortugas Laboratory, especially since the overall quality of the published works was exemplary.

Volume 1 of *Papers* reflected both the multidisciplinary nature of biological research being carried out at the Tortugas Laboratory and the caliber of investigator that Alfred Mayor had been able to attract. Included were reports on developmental and cellular biology, morphology, systematics, physiology, ecology, and parasitology. William Keith Brooks, of Johns Hopkins University, one of America's most respected zoologists of the day, contributed research on the morphology of several pelagic tunicates, commonly called salps, from the Gulf Stream. Also included in Brooks's paper was the description of a new species of salp, *Oikopleura tortugensis*. A second publication by Brooks, with coauthor Bartgis McGlone, explored the origin of the lung in a land snail. Accounts of the cytology and embryology of several sea stars, brittle stars, and sea urchins (Echinodermata) as well as a species of walking stick (Insecta) were given in three papers by Harvey E. Jordan, of the University of Virginia. H. F. Perkins, of the University of Vermont, described three new species of Hydrozoa, two of them (*Cladonema mayeri* and *Campanularia macrotheca*) from the moat surrounding Fort Jefferson. Edwin Linton, of Washington and Jefferson College, summarized investigations on parasitic worms of fishes, turtles, and shrimp at the Tortugas, while Charles H. Edmondson, of the University of Iowa, gave a brief description of a peculiar flagellate protozoan from the Fort Jefferson moat. Mayor contributed two papers, one on the reproductive ecology of the Atlantic palolo worm and another on the physiological basis of rhythmical pulsation in the scyphozoan jellyfish *Cassiopea*. He also added a note in tribute to Brooks, who died before the volume was published. Declared Mayor, "there lives in our hearts a love for this kindly man of culture, the modest, hopeful teacher, who was free from all trace of pedantry."[47]

The second volume of *Papers* also varied in content and included articles by several well-known American biologists. Investigations related to ecology included those by Rheinart P. Cowles, of Johns Hopkins, on the ghost crab *Ocypoda arenaria* (=*Ocypode quadrata*) and by Charles Stockard on the walking stick insect *Aplopus mayeri*. Papers on developmental biology were contributed by Stockard (rates of regeneration in the scyphomedusa *Cassiopea*) and by Charles Zeleny, of Indiana University (regeneration of the claws of the gulfweed crab *Planes minutus*). Edwin G. Conklin, then of the University of Pennsylvania, published two life-cycle studies. The subject of one of Conklin's inquiries dealt with swarming in, and development of, the thimble-shaped scyphomedusa *Linerges mercurius* (=*Linuche unguiculata*), and the other described two peculiar larvae of sea anemones (Actiniaria) found in the plankton at the Tortugas. Reports on animal behavior included a study by John B. Watson on noddy and sooty terns at Tortugas and another by Jacob E. Reighard on coloration of reef fishes. Frank Chapman contributed an article on life history observations of the brown booby, *Sula leucogastra* (=*S. leucogaster*), and the magnificent frigate bird, *Fregata aquila* (=*F. magnificens*).[48] Clearly, high standards of scientific content and production quality had been established from the outset in the CIW periodical on marine biology. The journal not only advanced scientific knowledge but also brought credit to the publication series, to the Tortugas Laboratory, and to Mayor.

Despite his thoughts of moving the laboratory away from its isolated location, Mayor continued to praise its advantages and achievements. For *Popular Science Monthly,* then a respected forum for scientists wishing to keep the general reader informed, Mayor drafted an article on "The Research Work of the Tortugas Laboratory," which appeared in print in April 1910. Including twelve of Mayor's excellent photographs of the Tortugas grounds, buildings, and fleet, the article provided an appealing bit of publicity for the laboratory. It stressed the aim of conducting modern biological research. "Fifty years ago Darwin

changed biology," Mayor said, "but it is only recently that it [biology] has exhibited decisive evidence of passing out of the qualitative into the quantitative stage
of its development." The newer approaches, he observed, were paramount in the
work of most of the biologists at Tortugas. Although systematics retained an
important place, he said, more was being done in the areas of physiology, embryology, and ecology.[49]

In Mayor's judgment, the major studies done at Tortugas included those by
Reighard on tropical fishes and by Watson on the nesting behavior and migra-
tory patterns of terns inhabiting the area. Watson had, of course, done research
at the Tortugas Laboratory in 1907, and Mayor liked his work so much that he
not only invited him to return in 1912 and 1913 but also persuaded Woodward
to award stipends to the man who later became known for his work as a behavioral psychologist. He viewed Watson as a victim of the summer-school teaching trap. Mayor also liked the work of Reighard on tropical fishes and that of
Stockard on *Cassiopea*. The greatest praise, however, went to Vaughan for his
research on corals. In Mayor's view, Vaughan's studies represented pioneering
work on the growth rate of corals and the influence of environmental conditions
on the morphological development of varying forms within the same species of
corals. The superb investigators who came to study at Tortugas, observed
Mayor, had been "undeterred by the roasting heat of desert sands . . . [and] the
ceaseless shrieks of sea-gulls." While the success of the laboratory was modest so
far, concluded Mayor, its promise was significant. Mayor also played upon
American pride, saying that the laboratory researchers were conducting studies
that would "lead to the discovery of laws of nature" and that they constituted a
"national effort."[50] One would not have suspected that Mayor had, off and on,
considered relocating the laboratory, but, then, he remained uncertain himself:
Tortugas was both ideal in some ways and undesirable in others as the location
of a marine biological station.

In early 1910, as he prepared for the sixth season of operating the Tortugas
Laboratory, Mayor was both disheartened and encouraged, and as the year pro-
gressed, he found himself alternating between the disparate moods. At the end
of March he received word that Alexander Agassiz had died while aboard a ship
returning from England. Mayor was among those who met the ship when it
docked in New York on April 2. Invited later to write a sketch of the life and
contributions of Agassiz for *Popular Science Monthly*, Mayor readily accepted.
Following his own feelings toward this giant in marine zoology, Mayor wrote
candidly of the man who had played such a large role in his own life and
career. After giving a sketch of the life of Agassiz, Mayor alternated between
praise and criticism, though he placed the latter in the context of the historical

development of biology and thus avoided any hint of condemnation. His assessment was keen but kind.[51]

Mayor praised Agassiz for his scientific studies and his leadership in promoting science. He also noted his personal obligation to his longtime associate: "when I was young and struggling, his hand befriended me, and his great mind gave direction to the thought of the life I have led." He added, "I think upon his spirit with gratitude and reverence, for he was my master in science." Among the criticisms offered by Mayor was the resistance of Agassiz to work in histology, physiology, and experimentation as a result of his viewing them, in Mayor's words, "as beyond the scope of zoology." Mayor also observed that Agassiz had never participated "in that stirring discussion of Darwinism which engrossed the attention of all of his contemporaries." In Mayor's view, Agassiz probably accepted the theory of evolution but held little interest "in the speculative side of science." The scientific expeditions made by Agassiz were important, noted Mayor, and he was a "colossal leader of a great enterprise." He added, however, that Agassiz was an "autocrat" on those voyages. Comparing him to Otto von Bismarck, Mayor characterized Agassiz as "fearless, resolute, quick to anger, definitely purposeful, and full of resource."[52] In any case, Mayor indicated no mean spirit in his comments; he simply pictured Agassiz as he knew him—a man to be admired and respected but a man whose authoritarian style revealed his sense of mission.

The book planned by Agassiz and Mayor had finally become solely the work of the latter, and its completion in 1910 came too late for the former to see. It is certain that Agassiz would have admired it—and perhaps even that he would have wished he had not withdrawn from the project. Agassiz had seen what Mayor could do as a systematist, a writer, and an artist. Mayor genuinely appreciated his mentor's considerable, if somewhat dated, knowledge of the medusae and his initial encouragement in the project. Although he had slowed the project and sometimes made matters difficult for Mayor, Agassiz had finally given it over entirely to the person who could do it best. In the long run Agassiz did Mayor a favor, for if *Medusae of the World* had appeared when it was originally intended to, it would not have been as thorough as it came to be.

Before he departed for Tortugas in April 1910, Mayor went to Princeton University to present a lecture, apparently at the joint invitation of two of its faculty members, Edwin G. Conklin and Stewart Paton. Conklin, one of America's most prominent biologists, had, of course, worked with Mayor at Tortugas during the summers of 1905 and 1907. The Mayors had known Paton and his wife since 1907, when all of them were in Naples, Italy. Paton was especially eager to get a part-time appointment for Mayor at Princeton, and he ultimately succeeded. Princeton's president notified Mayor of the appointment in July 1910. Meanwhile, Mayor had already decided to move Harriet and the children to the town of Princeton. To Alfred and Harriet, the move was important since it would result in good schools for their children and bring them close to special friends. The move also promised to ease Alfred's worries about his beloved Harriet, who was by then frequently fatigued and showing sure symptoms of tuberculosis. Harriet was also pregnant for the fifth time. She gave birth to a daughter, named Barbara, on August 14, 1910. In the meantime, the industrious mother had completed sculpting, in marble, a fine bust of her sister Anna.[53]

Fourteen scientists joined Mayor in 1910 for research at Tortugas, arriving and leaving at various times between May 2 and July 20. Among those coming for the first time was the Canadian-born ichthyologist William Harding Longley, of Goucher College in Baltimore, Maryland. Longley would return repeatedly, even after Mayor's death, and conduct significant studies of tropical fishes. Given the number and overall excellence of the researchers who worked at Tortugas during the season of 1910, Mayor had reason to hope for stellar success. Indeed, the investigators lived up to his expectations, but difficulties continued. Among the problems were "a plague of roaches," which devoured Mayor's collars and cuffs and gnawed off the backs of his books, and the loss of a load of earth when the schooner carrying it ran aground on a coral reef and was forced to dump the soil or sink. In early July one of the researchers, Aaron L. Treadwell of Vassar College, developed acute appendicitis, forcing Mayor to move him to Key West, a trip that took fifteen hours because of "a heavy sea and high wind with blinding sheets of rain." Only a few days later the Bryn Mawr College biologist David H. Tennent received word of the death of his young son, and Mayor was compelled to take him too to Key West. "We have had more kinds of bad luck than ever before," Mayor wrote to Harriet on July 12. More bad luck came long after the season had ended when, on October 17, a hurricane struck Loggerhead and badly damaged the laboratory. The storm destroyed the sailors' quarters, blew away part of the roof of the laboratory, knocked the machine shop from its foundation, toppled the windmill, and shredded leaves from virtually every plant on the key. Mayor was forced to make a special request of four thousand dollars to repair the battered facilities.[54]

Earlier, at the end of May in 1910, CIW president Woodward made a visit to the Tortugas Laboratory. Mayor met Woodward in Key West and transported him to Loggerhead Key in the *Physalia*. During the trip to the island, the *Physalia* encountered "a good head swell," and the event gave Mayor an opportunity to tell Woodward that the yacht was inadequate. He had suggested to Woodward as early as 1907 that the yacht was too expensive to operate and ought to be sold. From subsequent developments, however, it appears that

Mayor was hoping to obtain a larger and much-improved vessel. Woodward "was very enthusiastic" over the station, and as Mayor learned shortly over a month later, the CIW president had already arranged for the laboratory to obtain a new yacht, at a cost of twenty-five thousand dollars, more than four times the amount paid for the *Physalia* only six years earlier. Mayor was delighted to have the news, but he was irritated over another action that Woodward took without discussing the matter with him. "Dr. Woodward has been busying himself suggesting to the Trustees a stone building opened all year round at Tortugas," Mayor wrote to Harriet on July 12, and he added, "Just what I don't want." Mayor quickly informed Woodward that he was strongly opposed to the idea, but he told Harriet that he was concerned, as Woodward could be "hopelessly obstinate . . . when he gets a crazy idea in his head."[55] No doubt Woodward should have conferred with Mayor about his plan, but he obviously thought that he was doing him and the laboratory a great favor in proposing construction of a much more substantial stone edifice. After all, it was Mayor who had campaigned for a laboratory at Tortugas, and the Carnegie Institution had a right to build on the plan first laid out by Mayor, which included a provision that ultimately the laboratory would operate year-round.

In fact, Woodward had painted a rosy picture of the laboratory in a letter to CIW trustee Charles Doolittle Walcott on June 18, 1910. "The location of the laboratory," he said, "is extremely favorable" and "is also one of the most salubrious that can be found." In Woodward's estimation, the Tortugas station could, by "some additional degree of enlargement and permanence," become "one of the most important [laboratories] for the future of marine biology." Indeed, he maintained that it could rise to the status of "a world station." To do so, however, the Tortugas Laboratory would have to operate year-round, employ two or three "eminent investigators" in addition to Mayor, and acquire a better research vessel. Although Mayor had not recommended the first two measures, he had clearly convinced Woodward of the need for a yacht to replace the *Physalia,* which, noted Woodward to Walcott, was in a state of "rapid deterioration." The CIW president praised Mayor as "one of our ablest men," a leader who could shape the Tortugas Laboratory into a station "as famous . . . as the laboratory in Naples."[56] Mayor might complain of Woodward's case for permanent buildings on Loggerhead Key, but he could hardly grouse over the unstinting support of the CIW's leader.

While he praised the achievements of the laboratory in his report to the trustees for 1910, Mayor noted that the facility could not be opened during winters because the trade winds made collecting impossible, "larval forms" were absent, and college and university professors could not come then. Of course, the hurricane season kept the station closed during the late summer and the fall,

a point reinforced by Mayor's reference to the damage done by the severe storm
of October 17. One consequence of this storm was alteration of the shoreline
near the dock, which ultimately forced Mayor to extend the dock into deeper
water. Another station, one that could be operated during the winters, would be
the solution, noted Mayor, and he suggested Miami, Nassau, Montego Bay
(Jamaica), and New Providence (Bahamas) as possible sites. Still, despite problems, Mayor wanted to keep a laboratory at Tortugas, though he advised the
trustees that "a progressive change . . . in marine biology" necessitated "facilities
for long and elaborate studies," including more experimentation.[57] If Woodward
and the CIW trustees were somewhat confused by the vacillations of their
marine laboratory director, they remained supportive and continued to pour
money into the operation of the station.

*Katherine Goldsborough Mayer,
ca. 1866. From the Brantz
and Ana Mayor Collection.
Courtesy of Brantz and Ana
Mayor*

*Alfred Marshall Mayer,
ca. 1866. From the Brantz
and Ana Mayor Collection.
Courtesy of Brantz and Ana
Mayor*

*Alfred Goldsborough Mayor,
early 1870s. From the Brantz
and Ana Mayor Collection.
Courtesy of Brantz and
Ana Mayor*

*Alfred Goldsborough Mayor,
ca. age fourteen or fifteen.
From the Brantz and Ana Mayor
Collection. Courtesy of Brantz
and Ana Mayor*

*Alfred Goldsborough Mayor,
June 1889. From the Brantz
and Ana Mayor Collection.
Courtesy of Brantz and Ana
Mayor*

*Alfred Goldsborough
Mayor, ca. 1919.
From the Brantz and
Ana Mayor Collection.
Courtesy of Brantz and
Ana Mayor*

*Harriet Randolph Hyatt, 1890s. From the Brantz and Ana Mayor Collection. Courtesy of Brantz and Ana Mayor*

*Harriet Hyatt Mayor and three of her four children, August 12, 1913 (left to right: Brantz, Barbara, and Katherine). From the Brantz and Ana Mayor Collection. Courtesy of Brantz and Ana Mayor*

*Tortugas Laboratory, July 28, 1904. From the Brantz and Ana Mayor Collection. Courtesy of Brantz and Ana Mayor*

*Tortugas Laboratory, 1905. Courtesy of the Carnegie Institution of Washington*

*Sketch of Tortugas Laboratory, ca. 1905 by A. G. Mayor. Courtesy of the Carnegie Institution of Washington*

*Interior of Tortugas Laboratory, 1916. From the Brantz and Ana Mayor Collection. Courtesy of Brantz and Ana Mayor*

*The yacht* Physalia *of the Carnegie Institution of Washington, 1904. Courtesy of the Carnegie Institution of Washington*

*The yacht* Physalia *at the dock of the Tortugas Laboratory, July 1906. Courtesy of the Carnegie Institution of Washington*

*The vessel* Anton Dohrn *of the Carnegie Institution of Washington, 1916. Courtesy of the Carnegie Institution of Washington*

*Alexander Agassiz. Courtesy of the Smithsonian Institution Archives*

*Alpheus Hyatt. Courtesy of the Smithsonian Institution Archives*

*T. Wayland Vaughan. From the archives of the Scripps Institution of Oceanography.*

*William Harding Longley,
ca. 1926. Courtesy of the
Carnegie Institution of
Washington*

*Murray Island natives, 1913.
Photograph by Alfred Goldsborough Mayor. Courtesy
of the Carnegie Institution
of Washington*

APRIL, 1917 – 10½ lbs., 25½ INCHES IN CIRCUMFERENCE.

JULY, 1918 – 17⅛ lbs., 32¼ INCHES IN CIRCUMFERENCE.

*Coral transplanted by Alfred Goldsborough Mayor, 1919–20. From Alfred Goldsborough Mayor, "Growth-Rate of Samoan Corals,"* Papers from the Department of Marine Biology of the Carnegie Institution of Washington *19 (1924): pl. 15, fig. 52*

*Alfred Goldsborough Mayor, ca. 1918. From* Papers from the Department of Marine Biology of the Carnegie Institution of Washington *19 (1924): insert*

# Monographs on Medusae

After spending almost twenty years of his life collecting, studying, describing, and illustrating jellyfishes, Mayor was ready to share the results of his research in a major monograph. The culmination of his efforts was a magnificent three-volume work entitled *Medusae of the World,* published in 1910 by the Carnegie Institution of Washington.[1] Medusae, or jellyfishes, an abundant, ubiquitous, and ecologically important group of invertebrates in the sea, were poorly known in Mayor's day. The objectives of these volumes were to advance knowledge of them, to bring together in a single work existing knowledge on all known jelly-fish species worldwide, and to propose a new and improved classification system for the group.

Mayor had commenced work on the project in 1892, at the outset of his career in zoology, while a student at Alexander Agassiz's marine laboratory in Newport, Rhode Island. Indeed, it was Agassiz who initially proposed the study, although the original objectives were modest compared to the content of the volumes that appeared nearly two decades later. Agassiz had proposed that Mayor assist him in characterizing the medusae, siphonophores, and ctenophores of the Atlantic coast of North America. To attain their goal, Agassiz had sent Mayor on research trips to Halifax, Nova Scotia; to Eastport, Maine; to Charleston, South Carolina; and to the Dry Tortugas, Florida. At each of those locations Mayor undertook to collect medusae, to describe the species he obtained, and to prepare color illustrations of them. Observations on species found at those locations supplemented work that he undertook elsewhere along the coast, including Newport and vicinity, the Woods Hole region in Massachusetts, and the Bahamas, during a cruise with Agassiz on the steamer *Wild Duck.*[2]

By 1900 Mayor had completed accounts of the medusae of the North American East Coast without any noteworthy contribution from his overcommitted mentor. Four years later, with numerous other interests and responsibilities still having priority, Agassiz had returned the languishing manuscript to Mayor and instructed him to proceed with it as he saw fit. By this time, however, a considerable amount of new information had been generated on species of the American Atlantic. In addition Mayor had benefited from opportunities to collect and study medusae around the globe during scientific expeditions undertaken with Agassiz. With revision of his manuscript a necessity in any case, Mayor resolved to broaden its scope to encompass the medusae of the world. Having greatly expanded its coverage, both in geographic area and in number of species, Mayor had to undertake several years of additional research and writing before he could complete a new manuscript.[3]

It is clear that Mayor became captivated by his subject animals, for he wrote that one could not "remain insensible to the rare grace of form and delicate beauty of color of these creatures of the sea," and he wrote of "the absorbing study of the medusae."[4] Intellectually challenged by his studies of them, Mayor observed that "with the exception of the sponges and corals, there is no phylum of the animal kingdom more difficult to classify than the medusae."[5] Research on these marvelous "creatures of the sea" unquestionably provided him with many hours of discovery, excitement, awe, inspiration, and happiness.[6] Such delightful times would nevertheless have been interspersed with periods of tedium and drudgery during preparation of dry descriptions and lists of scientific names and literature references relevant to each species. Mayor's diligence and persistence resulted in an encyclopedic work that was well written and beautifully illustrated.

The three volumes of quarto-size books comprise 735 pages of text, 76 color plates, and 428 text figures. Printed on high-quality paper, the pages and plates remain in good condition more than nine decades later. Although Mayor was somewhat prone to spelling mistakes, the volumes are relatively free of misspellings and typographical errors. For the record, and of importance in zoological nomenclature, the official date of publication of volumes 1 and 2 was August 9, 1910, while that of volume 3 was August 25, 1910.[7] Mayor expressed regrets in the introduction to the third volume that Agassiz did not live to see in print the work that he had inspired two decades earlier, for the Harvard naturalist had died less than five months before it was published.[8]

In addition to providing broad geographic coverage in *Medusae of the World,* Mayor sought to make the three volumes monographic in scientific scope and content. When such was known, he noted the colors of living specimens, provided information on abundances and dates of collection, and described known variations within a species. Yet the volumes were more than a compendium on the systematics of the group. Mayor also summarized information published on other biological aspects of the species studied, including life cycles, embryology, cytology, ecology, physiology, regeneration, and behavior.

Superb illustrations accompanied the text in all three volumes. Mayor prepared nearly all of the figures in the color plates, and he drew and painted them based on examinations of living specimens. Thus, the shapes and colors of medusae portrayed in these figures accurately reflect the actual appearance of these animals when alive. Mayor's artistic ability was almost certainly a factor in Agassiz's decision to assign the medusa project in 1892 to his new and then unproven student. The decision proved to be a fortunate one for natural history and marine biology, and especially for students of medusae who followed. Most of the text figures, from the widely scattered works of many other authors, were done by Carl Kellner, a zoologist and artist employed by Mayor at the Tortugas Laboratory of the Carnegie Institution of Washington. Included as well among the text figures and plate figures are several works by William Keith Brooks, a major American naturalist associated with the Johns Hopkins University.

The first two volumes of Mayor's trilogy on medusae dealt with the hydromedusae, or "veiled medusae" (phylum Cnidaria, class Hydrozoa). These are mostly small jellyfishes distinguished in part by the presence of veils or "vela" on their bells. The velum enhances swimming speed and facilitates agility by regulating the angle that water is forced out of the bell cavity of the medusa. While 850 species of hydromedusae are known to science today,[9] Mayor's two volumes on hydromedusae included formal accounts of some 565 species or forms of species that were recognized at that time, with brief mention of others that he regarded as being of highly questionable status. In his classification system, Mayor assigned these species to 147 genera. He provided information on the polyp, or on various larval stages, of nearly one-quarter of the species, and he illustrated more than half of them. In all, Mayor established new names of two genera and ten species of hydromedusae in these two volumes.

Mayor devoted the third volume of *Medusae of the World* to the scyphomedusae, or "true jellyfishes" (phylum Cnidaria, class Scyphozoa). This group currently comprises some 180 species, with familiar ones being the moon jelly (*Aurelia aurita*), the sea nettle (*Chrysaora quinquecirrha*), the lion's mane jelly (*Cyanea capillata*), the cannonball jelly (*Stomolophus meleagris*), and various upside-down jellies (*Cassiopea* spp.). Volume 3 also included accounts of the cubomedusae or box jellies, distinguished in part by their cuboidal bells. Box jellies are now split from the Scyphozoa as a distinct class (phylum Cnidaria, class Cubozoa). The bottom-dwelling and attached Stauromedusae, or stalked jellyfishes, long regarded as aberrant scyphozoans, have likewise been recognized recently as a separate class of Cnidaria (class Staurozoa).[10] Mayor also included brief accounts of eighteen species of fossil medusae. He ended the book with an appendix updating scientific names to accord with the International Code of Zoological Nomenclature and providing new information on hydromedusae and scyphomedusae, *sensu lato*. Included in his coverage of the Scyphozoa, as then understood, were 207 species and varieties or forms of species, assigned to seventy genera. Of these, 18 were box jellies, in six genera (now referable to the class Cubozoa); 30 were stalked jellies, in twelve genera (now class Staurozoa); and the remaining 159, in fifty-two genera, were scyphozoan jellyfishes (class Scyphozoa). Mayor established the names of two new genera and six new species in the third volume, and he illustrated 90 of the 207 kinds of jellyfishes discussed in the work.

Contemporaries in cnidarian research immediately recognized Mayor's magnum opus as an important milestone in the advancement of knowledge in that area. In a two-page review of volumes 1 and 2 appearing in the October 28, 1910, issue of the prestigious American journal *Science,* Charles Cleveland Nutting, noted for an exceptional series of monographs of his own on hydroids, wrote: "'Medusae of the World' is a monumental work which will take the very first rank and be a classic of which the Carnegie Institution may well be proud, and for which the author is to be heartily congratulated." Nutting praised the exhaustive coverage of the work and found it "replete with interesting facts." He commented favorably on the illustrations and noted that placement of the color plates adjacent to the relevant text, a practice usually rejected as inconvenient by publishers, would be welcomed by users. Nutting offered only three criticisms of any note. First, he found the font size used in synonymy lists to be too small and hard on the eyes when these were studied over an extended period. Second, he thought that Mayor should have provided more discussion in support of his ideas on evolutionary relationships among orders of hydromedusae. Third, Nutting contended that some discussion of medusa morphology and terminology at the outset of the work would have been helpful to those unfamiliar with the group. Yet his admiration for the treatise left Nutting "in no mood for criticism of small details."[11] Nutting clearly understood that this was an exceptional work by a competent and dedicated marine biologist.

Later in 1910 another review of *Medusae of the World* appeared in the equally prestigious British journal *Nature.* While the anonymous reviewer was overwhelmingly positive in his assessment, he offered two criticisms of the work. First, he suggested that a table outlining the respective classifications of hydroids and medusae would have been instructive. He added, however, that "it seems somewhat ungracious to begin a notice of work that is characterized by so many excellent features by complaining about an omission." Second, the reviewer expressed regret that Mayor had adopted certain changes in scientific names, even though mandated by the International Code of Zoological Nomenclature. For example, he did not like the use of the generic name *Craspedacusta* in place of *Limnocodium* for the well-known species of freshwater medusa. Notably, however, *Craspedacusta* has now been universally adopted, with little confusion or difficulty, as the valid name of the genus. On the whole, however, the reviewer was pleased with the volumes and commended Mayor for his multidisciplinary coverage and for his illustrations, particularly the color plates. "It would be difficult," he said, "to express adequately our admiration of the seventy-six coloured plates with which this monograph is illustrated." In addition the reviewer praised Mayor's approach in carefully reviewing the characters used to differentiate certain questionably defined species and then offering his own opinion on their status. He ended his review by saying, "We cannot close this

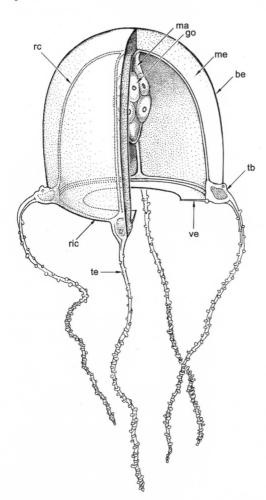

*Diagram of a hydromedusa, with part of bell cut away. be = bell, go = gonad, ma = manubrium, me = mesoglea ("jelly"), rc = radial canal, ric = ring canal, tb = tentacle bulb, te = tentacle, ve = velum*

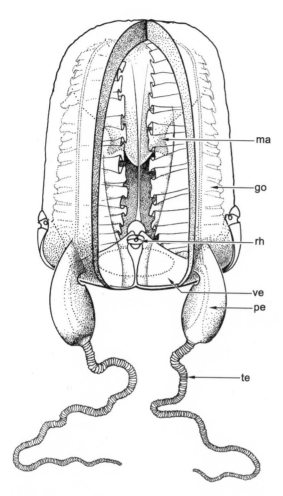

*Diagram of a cubomedusa (*Tamoya *sp.), with part of bell cut away.* go = *gonad,* ma = *manubrium,* pe = *pedalium,* ve = *velarium,* rh = *rhopalium,* te = *tentacle*

notice without again expressing our thanks to Mr. Mayer for his most magnificent and serviceable memoir. It is really a great work, and will mark a great step of progress in the literature of the subject."[12] Such praise could have given Mayor cause to boast, but in his annual report to the Carnegie Institution of Washington he made little fanfare, modestly mentioning the volumes only in a list of papers published during the year on research undertaken at Tortugas.[13]

In a letter to Mayor dated April 15, 1912, the eminent German biologist and medusologist Ernst Heinrich Philipp August Haeckel expressed admiration for Mayor's work. Generous in his praise, Haeckel complimented Mayor for his "successful studies." He also expressed appreciation of Mayor's "fine artistic talent" and lauded "the many marvelous illustrations of living medusae" in the volumes. Haeckel noted that he was especially impressed by the "perfection" of the drawings. In addition he indicated that he was "pleased" that Mayor had

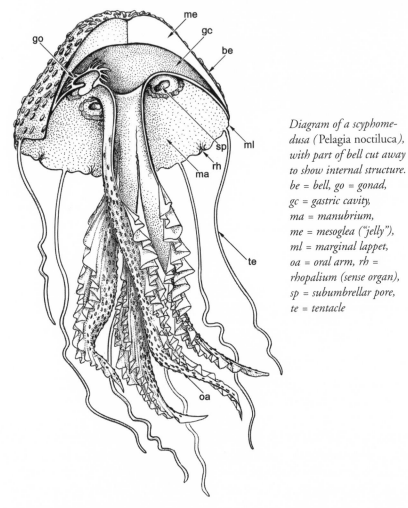

*Diagram of a scyphome-
dusa (*Pelagia noctiluca*),
with part of bell cut away
to show internal structure.
be = bell, go = gonad,
gc = gastric cavity,
ma = manubrium,
me = mesoglea ("jelly"),
ml = marginal lappet,
oa = oral arm, rh =
rhopalium (sense organ),
sp = subumbrellar pore,
te = tentacle*

corrected "many errors" in his own works on medusae published three decades earlier.[14]

Certain publications or series of publications on medusae over the centuries may be ranked as classic works because of their exceptional and lasting scientific value. Mayor's three-volume treatise beyond a doubt qualifies as one of these. The relative importance of this work becomes more evident when placed in company with four others of essentially similar rank and comprehensiveness. While these are difficult to compare with Mayor's tomes, and with one another, because of differences in objectives, approach, and content, each represents a significant milestone in medusa research.

The first of these is a lengthy pair of works by Ernst Haeckel. Haeckel's first volume was a monograph on hydromedusae published in 1879. The second was

A. G. Mayer del.

B. Meisel. lith.

*Illustrations of the Tahiti land snail* Partula. *From* Some Species of Partula from Tahiti. A Study in Variation. Memoirs of the Museum of Comparative Zoölogy, at Harvard College *26 (1902): unnumbered plate. (1, 2)* Partula hyalina; *(3–8)* Partula otaheitana; *(9, 10)* Partula filosa; *(11–13)* Partula sinistrorsa; *(14, 15)* Partula nodosa *var.* sinistralis

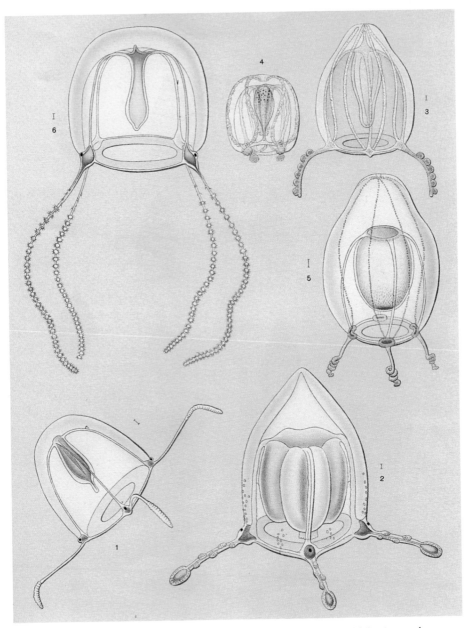

*Plate 5 from* Medusae of the World: *six species of anthoathecate medusae. (1)* Sarsia angulata *(now* Coryne angulata*) from Nassau, Bahamas; (2)* Corynitis agassizii *(now* Sphaerocoryne agassizii*) from Charleston, South Carolina; (3)* Ectopleura minerva *from Tortugas, Florida; (4, 5)* Ectopleura dumortierii *from Newport, Rhode Island; (6)* Sarsia mirabilis var. reticulata *(status uncertain) from Nahant, Massachusetts*

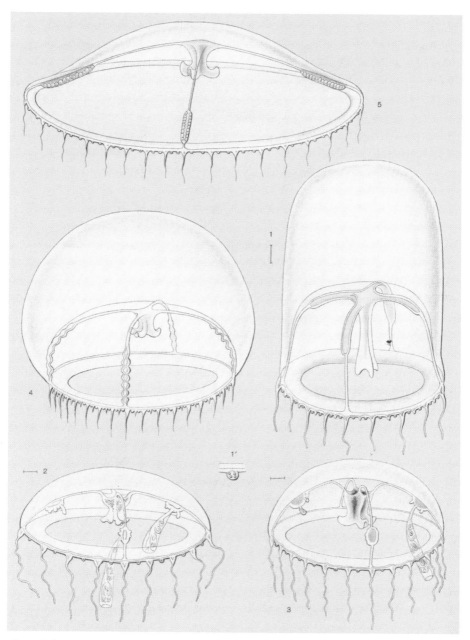

*Plate 34 from* Medusae of the World: *four species of leptothecate medusae. (1)* Phialidium gelatinosum *(now* Clytia gelatinosa*) from Tortugas, Florida; (2, 3)* Phialidium mccradyi *(now* Clytia mccradyi*) from Nassau, Bahamas, and Tortugas, Florida, respectively; (4)* Phialidium globosum *(now* Clytia globosa*) from Tortugas, Florida; (5)* Phialidium languidum *(now* Clytia languida*) from Tortugas, Florida*

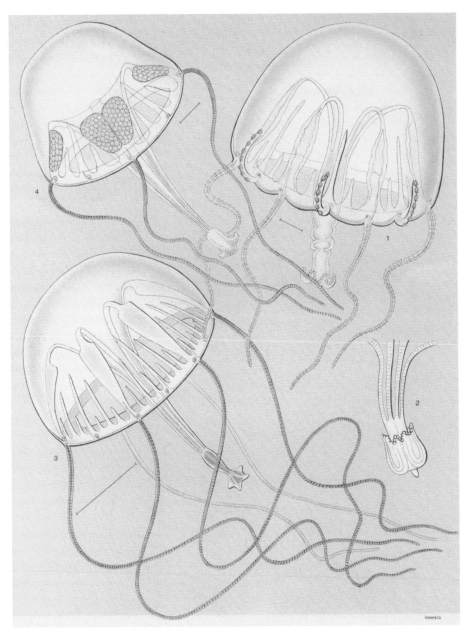

*Plate 53 from* Medusae of the World: *two species of Trachymedusae. (1–3)* Geryonia proboscidalis *from Tortugas, Florida; (4)* Liriope tetraphylla *from Tortugas, Florida*

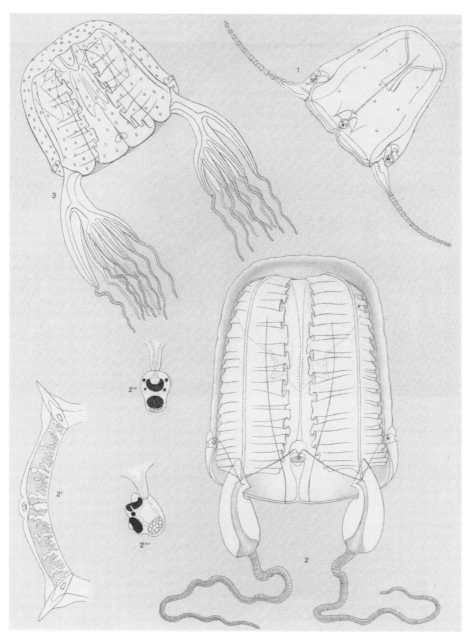

*Plate 57 from* Medusae of the World *(1910): three species of cubomedusae.* (1) Carybdea
xaymacana *from Nassau, Bahamas;* (2) Tamoya haplonema *from Long Island, New York;*
(3) Chiropsalmus quadrumanus *from Beaufort, North Carolina*

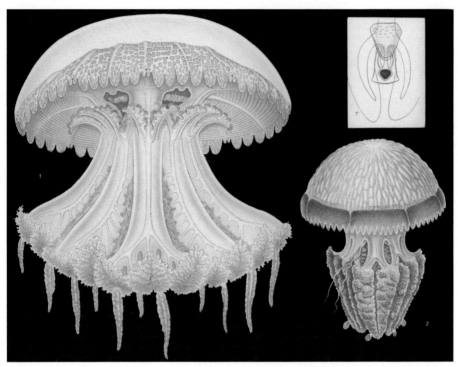

*Plate 74 from* Medusae of the World: *two species of scyphomedusae. (1)* Rhopilema verrilli *from Pamlico Sound, North Carolina; (2)* Crambione cookii *from Cooktown, Queensland, Australia*

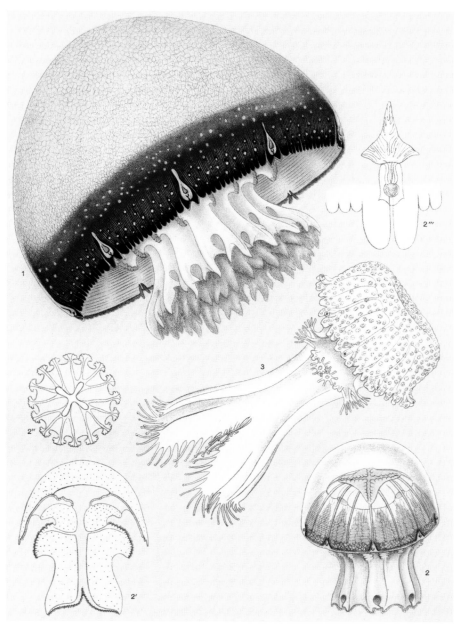

*Plate 75 from* Medusae of the World: *the scyphomedusa* Stomolophus meleagris. *(1) adult from Fernandina, Florida; (2, 3) young medusae from Charleston, South Carolina; (2', 2", 2''') certain anatomical details of the juvenile shown in fig. 2*

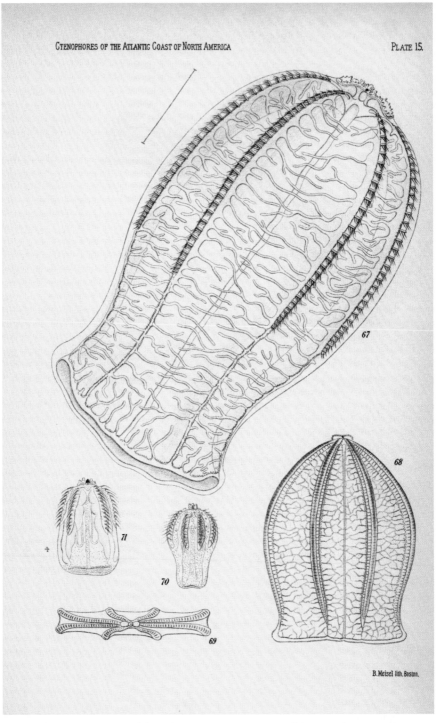

*Plate 15 from* Ctenophores of the Atlantic Coast: *two species of the ctenophore genus* Beroe. *(67)* Beroe cucumis *from Eastport, Maine; (68, 69)* Beroe ovata *from St. Mary's River, Maryland; (70, 71) larval stages of* Beroe *sp. from Newport, Rhode Island*

a companion volume on scyphomedusae (including stauromedusae and cubo-medusae) published in 1880.[15] Haeckel energetically sought to discover, describe, and name as many kinds of medusae as possible. No other author even approaches him in the number of new species of medusae that he established, and he is credited as author of 134 of nearly 775 species and varieties of medusae treated in Mayor's three volumes. Nevertheless, some criticisms have been leveled at Haeckel's work, including the use of minute and sometimes meaningless differences in specimens to discriminate supposed species. Many of these so-called species are now known to have been based simply on growth stages, races, or variations of a given species, or on artifacts of preservation, and have been abandoned or "sunk in synonymy." Moreover, while Haeckel prepared elaborate and striking illustrations to accompany morphological descriptions of his medusae, his drawings often represented flamboyant re-creations of what he imagined a given medusa would look like in life rather than portraying the actual form of the specimen before him. Mayor dismissed Haeckel's ideas on classification as "too artificial to accord with nature" and observed that many of his genera were "founded upon intergrading characters, and are thus imperfectly separated."[16] In spite of all these deficiencies, the iconoclastic Haeckel must be credited with advancing knowledge on biodiversity of medusae to an extent unmatched by any other scientist.

Another set of references approaching the worldwide geographic scope and general taxonomic utility of Mayor's volumes is that of the Danish marine biologist and noted medusa specialist Paul Lessenius Kramp. Like Mayor, his research culminated in three widely used reference works on these invertebrate animals: a volume on hydromedusae of the Atlantic Ocean and adjacent waters (1959); a descriptive synopsis of the hydromedusae and scyphomedusae (including cubomedusae and stauromedusae) of the world (1961); and a report on the hydromedusae of the Pacific and Indian oceans (1968).[17] Kramp acknowledged the value of Mayor's volumes in his publications, saying that "A. G. Mayer . . . provided us with a monograph which proved indispensable for all students of the group,"[18] and he used Mayor's volumes as his starting point.

In 1953 Sir Frederick Stratten Russell, former director of The Laboratory, Marine Biological Association of the United Kingdom, Plymouth, published a large book on hydromedusae and, in 1970, another one on scyphomedusae of the British Isles.[19] The content of these classic studies is, in fact, more comprehensive than the titles suggest. While he focused on species known from waters around the United Kingdom, Russell also reviewed much of the relevant literature on the biology of jellyfishes at the time. In the introductions of these volumes, he acknowledged the monographs of Mayor as milestones in research on medusae.

A final pair of volumes that approach the status of Mayor's monographs, dealing with Hydrozoa and Scyphozoa of the seas of the former USSR, came from the pen of Donat Vladimirovitch Naumov (1921–84) of St. Petersburg, Russia.[20] Naumov's volume on Hydrozoa, published in 1960, incorporated a unified classification scheme of hydroids and medusae, and advanced a theory that these two stages may evolve at different rates and in different directions. Translated into English by the Israel Program for Scientific Translations and published in 1969, it is still a widely used reference. Naumov's book on Scyphozoa, published in 1961, is an important reference for investigations on scyphomedusae and stauromedusae of northern seas. This work is also in Russian, although an English translation of the section on stauromedusae is apparently available for study at the Natural History Museum in London. Mayor's work was cited repeatedly in Naumov's two publications, especially in synonymy lists.

Unlike much of the literature of science, which at best may be considered "cutting-edge" or "ground-breaking" for a few weeks, months, or rarely years, these classic works remain as indispensable references on systematics and biology of jellyfishes. Although Mayor's monographs are now almost a century old, and while his classification system and scientific names are dated, they have never been completely replaced. They constitute one of the greatest and arguably the most comprehensive works overall ever written on the group.

In order to appreciate the challenges and difficulties that Mayor encountered in pursuing his research on jellyfishes, it is helpful to understand the complexities of these creatures that so fascinated him. Some people are likely to conjure up images of jellyfishes as large, amorphous, and repulsive gelatinous masses stranded on a beach, or perhaps as sinister blobs in the water to be avoided as a possible danger or public health hazard. Certain kinds are indeed venomous, but such species are relatively few in number and most are immediately recognizable. Some are indeed quite large, with a few reaching three feet (approximately one meter) or more in diameter as adults. The overwhelming majority of species, however, are tiny creatures having a diameter less than three-eighths of an inch (1.3 centimeters). Scarcely, if at all, visible to the naked eye in the water, these small jellyfishes must first be captured with a plankton net, a dip net, a bucket, or a bottle, and then detailed study of them must be made under a dissecting microscope. Close observation of any medusa reveals it to be a unique kind of creature, far more peculiar in body form and lifestyle than any of the fanciful extraterrestrial beings portrayed in most works of science fiction. Moreover, jellyfishes are graceful and beautiful aquatic creatures of varied shape, color, and habits.

Medusae belong to a major group or phylum of animals called Cnidaria, a name derived from the Greek word *"knide,"* meaning "nettle." Included with medusae in this phylum are corals, sea anemones, sea fans, hydroids, siphono-phores, and box jellies. As implied by the name Cnidaria, representatives of the group have developed a "weapons system" like no other in the animal kingdom: they produce microscopic stinging capsules (nematocysts) used in both food capture and defense. All cnidarians possess nematocysts, but most species are incapable of stinging humans to any noticeable extent because they are either too small or the toxins contained in their nematocysts are too weak to be felt. A few, however, including box jellies, fire corals, sea nettles, and the Portuguese man-of-war, are well known to be venomous. Still fewer, such as the deadly box jellyfish (*Chironex fleckeri*) of Australian waters, are capable of killing a healthy adult human in a matter of minutes.[21]

In terms of their body plan, cnidarians are relegated to two basic forms, polyp and medusa. The polypoid form is readily apparent in the sac-shaped bodies of sea anemones and corals. The medusoid form, typically resembling a thick gelatinous bell or umbrella, is seen in the familiar jellyfishes. Cnidarian polyps are usually sedentary, bottom-dwelling, and colonial (e.g., like most corals and hydroids), while their medusae are mostly free-swimming and soli-tary (e.g., jellyfishes). A given species of cnidarian may be strictly polypoid, or strictly medusoid, or its life cycle may alternate between the two stages. In his *Medusae of the World,* Mayor directed his attention mainly to the medusa stages of cnidarians, although he also provided information on polyps where these were known. He addressed virtually all of the species of medusae recognized by science at the time.

Medusae are moderately diverse animals biologically, with about 1,100 species of them currently being recognized worldwide. Of these, approximately 850 are hydromedusae (veiled medusae, class Hydrozoa), 180 are scyphome-dusae (true jellyfishes, class Scyphozoa), 25 are cubomedusae (box jellies, class Cubozoa), and 35 are stauromedusae (stalked jellyfishes, class Staurozoa). Not included in these numbers are some 150 species of hydrozoans called siphono-phores, a group which Mayor also studied, with the beautiful but venomous Portuguese man-of-war (*Physalia* spp.) being the most familiar representative. Jellyfishes are mostly marine creatures, ranging from polar seas to the tropics, and from shallow waters to the deep sea. Only a few of them, all hydrome-dusae, inhabit freshwaters. They are recognized as an important component of marine food webs, both as prey and predators.[22] Although quite varied in form, medusae are all relatively simple in basic anatomy. In essence, each resembles a bell, umbrella, or thimble, and most of them are hardly more than a few mil-limeters, or a small fraction of an inch, each in diameter. A jellylike substance (mesoglea) within this bell, and situated between the ectoderm (an external layer of cells, forming an epidermis) and the endoderm (a layer of cells forming a lining within internal cavities), affords both a kind of hydraulic skeleton for the animal and neutral buoyancy for it in the water. It is from this gelatinous mesoglea that the common name "jellyfish" is derived.

Threadlike, stringlike, or cordlike tentacles, densely armed with nematocysts and used in capturing prey and in defense, hang from the rim of the bell in a majority of species. While the bell and jelly of hydromedusae and cubomedusae are usually clear or translucent, the tentacles, gonads, and especially the manubrium, or mouth-stalk, and tentacle bulbs may be variously and beautifully colored. In scyphomedusae even the bell is commonly pigmented to some extent. Pigmentation tends to be particularly intense in deep-sea species of medusae, which often appear red or even black. Some species of jellyfishes harbor algal symbionts in their tissues, take on some of their coloration, and derive nutrients and oxygen from these hosted organisms. Well-known examples are brown- to green-colored species of upside-down jellies (*Cassiopea* spp.), studied extensively by Mayor and others at Tortugas.

From the underside of the bell of the medusa, and much like the clapper of a real bell, hangs a tubular or rectangular mouth-stalk, or manubrium, with a mouth at its end. In some scyphomedusae, such as the familiar sea nettles (*Chrysaora* spp.), long and frilly oral arms extend from the ends of the manubria like curtains. In others of the same group, the oral arms are fused into a gelatinous mass, as in the cannonball jelly (*Stomolophus meleagris*) and its relatives. Cubomedusae possess structures, called the velaria, that are analogous to the vela at the bases of the bells of hydromedusae. Mayor maintained that the resemblance of the two structures was the result of parallelism and that they were not homologous, or similarly related in structural origin.[23]

Mayor and other biologists have distinguished hundreds of species of medusae on the basis of differences in their morphology.[24] Yet, the relatively simple anatomy and variable morphology of medusae, the significant changes in appearance that they undergo during growth, and the artifacts introduced by preservation make them an exceedingly frustrating group for taxonomists seeking to discriminate species. In discussing the scyphozoan genus *Pelagia*, Mayor observed that "the 'species' are in almost hopeless confusion" and concluded, "As in *Cyanea* and *Aurelia* so in *Pelagia* we find that the Linnaean system is inadequate to express the relationship of the numerous, closely related forms." Later, in remarks on the related genus *Chrysaora*, he noted, "The forms of *Chrysaora* are so imperfectly separated one from another that were it not for the fact that many minute distinctions have been pointed out between them, I would greatly prefer to consider them all to be one variable species." He believed that "we have here only one species, the varieties of which defy classification in terms of the Linnaean system."[25]

Further complicating the classification and scientific nomenclature of these animals is the fact that many species undergo an alternation between two quite distinct morphological stages, the planktonic medusa and the bottom-dwelling polyp. Mayor addressed the crux of this problem in the introduction of volume 1 by noting, "Much confusion has been introduced through the habit, in vogue among marine expeditions, of sending all of the medusae to one specialist and the hydroids to another."[26] Thus, different scientific names have often been applied to the two stages of a given species, especially in the Hydrozoa. Moreover, different classification systems developed early on for medusae and hydroids.

Mayor recognized that dual nomenclature and separate classifications for polyps and medusae were entirely illogical. When he and others of his day discovered that two or more different scientific names had been applied to a single species, they solved the problem by applying the "Principle of Priority," by which the oldest name in most cases was adopted as the valid one. At that time, however, too little progress had been made to merge the two classification systems. Mayor noted that he had "endeavored to bring the systematic arrangement of the medusae more nearly into accord with that of the hydroids," but he concluded that "a strictly natural system including both hydroids and medusae can not be constructed, for many of the hydroids remain undetermined." He also observed that "dissimilar hydroids (*Syncoryne, Stauridium*) may give rise to similar medusae (*Sarsia*); or the reverse may be the case." Perceptively, Mayor noted, "These and many other cases of a similar nature interpose a barrier to our attempts to invent a natural system which includes all hydroids and medusae within its embrace."[27] Life-cycle studies over the last century have resolved many of these problems, and application of the provisions of the International Code of Zoological Nomenclature have helped too, but much yet remains to be done. A new tool that is helping to resolve some of the confusion in the systematics of medusae and their polyps is the application of molecular methods in determining differences in DNA.

Most medusae are able to swim, generally by a form of jet propulsion. Mayor was particularly taken with a description by H. F. Perkins of the swimming and feeding behavior of the hydromedusa *Gonionemus murbachii* from the Eel Pond at Woods Hole, Massachusetts. Thus, in the second volume of his *Medusae of the World* he quoted Perkins's account: "On cloudy days or toward nightfall the medusa is very active, swimming upward to the top of the water, and then floating back to the bottom. In swimming it propels itself upward with rhythmic pulsations of the bell-margin, the tentacles shortened and the bell very convex. Upon reaching the surface the creature keels over almost instantly, and floats downward with bell relaxed and inverted and the tentacles extended far out horizontally in a wide snare of stinging threads which carries certain destruction to creatures even larger than the jelly-fish itself."[28]

It is common knowledge that jellyfishes are mostly water. Mayor documented this by quoting research of C. F. W. Krukenberg, who found that the familiar moon jelly (*Aurelia aurita*) contained 95.34 percent water and 4.66 percent solids.[29] Moreover, some medusae are known to produce "living light" or bioluminescence, a few noteworthy examples being the oceanic jelly, *Pelagia noctiluca;*[30] hydromedusae of the genus *Aequorea;*[31] and hydroids of the medusa *Obelia pyriformis.* According to Mayor, a medusa of *Phortis pyramidalis* (=*Eirene pyramidalis*) from Florida and the Bahamas, when disturbed, "glows with an intense blue-green phosphorescence, far more brilliant than that of any other medusa we have observed."[32] Vulnerable though these fragile and often minuscule creatures may appear, they have been making their way successfully in the oceans feeding, respiring, reproducing, and otherwise responding to their environment for a much longer time than the human species has inhabited the earth.

Mayor was interested in medusae for purely academic reasons, seeking to study their diversity worldwide and to document their biology. Besides their academic appeal, however, jellyfishes exert an impact on humans in various ways, some negative and some positive. The most notorious of all the jellyfishes by far are the box jellies, several species of which (for example, the box jellyfish, *Chironex fleckeri,* and the Irukandji jellyfish, *Carukia barnesi*) are highly venomous and have been responsible for human fatalities. Certain scyphomedusae are also capable of severely stinging humans, well-known examples being various sea nettles and their relatives (*Chrysaora* spp., *Pelagia* spp., *Sanderia malayensis*), lion's mane jellies and related species (*Cyanea* spp., *Drymonema* spp.), and upside-down jellies (*Cassiopea* spp.).[33] Considerable interest in jellyfishes has arisen in recent years as a result of conspicuous population explosions of certain species and the negative impacts of these so-called blooms on a variety of human activities. Also of concern has been the negative impact of invasive species such as the Australian spotted jellyfish (*Phyllorhiza punctata*) outside their areas of origin.[34] In Mayor's day this medusa was known only from Australia, but it has now spread even into the Atlantic Ocean.

On a more positive note, dehydrated or pickled jellyfish is considered a delicacy in parts of Asia, including China, Taiwan, Korea, and Japan, and commercial fisheries exist for certain species (for example, *Rhopilema* spp., *Catostylus mosaicus, Stomolophus meleagris*). The annual harvest, most of it from China, has reached more than three hundred thousand tons in recent years.[35] Mayor alluded to the use of jellyfish as food in the third volume of his work on medusae. In discussing the edible jellyfish *Rhopilema esculenta,* he noted: "It is the custom in Japan to preserve it with a mixture of alum and salt or between the steamed leaves of a kind of oak. It is then soaked in water, flavored with condiments, and when so prepared constitutes an agreeable food." Mayor also reported being told that the box jelly *Chiropsalmus quadrigatus* was preserved in vinegar and eaten in the Philippines. In the same volume Mayor implied that cnidarians may be useful as indicators of the quality of their environment. Noting that the stauromedusa *Halimocyathus platypus* had not been seen since it was described from Chelsea Beach, Massachusetts, by Henry J. Clark in 1863, Mayor commented that "contamination of the sea-water in this region has destroyed the Stauromedusae which once abounded there, and which are now exceedingly rare along the entire New England coast."[36]

In the scientific content of his monographs, Mayor included insights into the group, including, for example, a statement that there was little evidence to suggest that scyphomedusae had evolved from hydromedusae. Instead, he held that the medusoid form in the two groups had likely been acquired independently. Of particular interest to Mayor were the analogies between the rhythmic pulsations of medusae and those of the hearts of vertebrates. He pursued this line of thought in greater detail as part of his physiological research on medusae, and he expressed hope that some practical application might ensue from such research. Inclined to believe that the stauromedusae were the most degenerate scyphozoans, he thought these sedentary medusae represented sexually mature polyps or scyphistomae. While this view prevailed until recently, stauromedusae are now held to be a distinct class of Cnidaria and not scyphozoans.[37]

Particularly irritating to Mayor was the tendency on the part of some cnidarian biologists to describe new species with insufficient justification. His lack of patience over the issue is apparent in his discussion of the hydroid polyps of the genus *Obelia*. "Undoubtedly many 'species' have been founded on mere geographical or environmental characters of a racial rather than specific nature," he wrote, and the result was that "hopeless confusion has . . . been introduced into the synonymy." Thus, Mayor offered a "description of the so-called species . . . [in] the hope that an inspection of the inextricable confusion here displayed may deter future students from further inflictions of such 'discoveries' upon the world." Added Mayor, sarcastically, "Every bushel basket full of *Obelia* hydroids, collected at random along our shore, is sure to contain several dozens of 'new species.'"[38]

Later in the same work Mayor expressed a similar opinion with respect to hydromedusae: "Many of our so-called species must be only geographical varieties or races of widely distributed forms, and the great advance in systematic studies of medusae will come in future from a comparison of reared specimens from numerous localities over the world." He said, "Such studies will probably reduce rather than increase the number of 'species' of medusae." Mayor repeated the same theme in his discussion of the Scyphozoa, saying, "It appears that the numerous so-called species of *Cyanea* intergrade to such a degree that we can not maintain them, and I believe there are only two species: *C. capillata* of the north temperate and Arctic regions and *C. annaskala* of the south temperate and Antarctic." Moreover, noted Mayor, "In common with *Pelagia, Chrysaora, Dactylometra, Aurelia,* and other world-wide forms of medusae, growth-stages, color varieties and local races have frequently been described as separate species, but as our knowledge increases many intergrading forms come to light thus reducing the so-called species to a few dominant types with numerous, closely related offshoots." It was unfortunate, Mayor said, that "the aim of the old systematic zoology was mainly toward the emphasizing of distinctions rather than the indication of affinities and the discovery of relationships."[39] Mayor's inclination toward being a "taxonomic lumper" would be disputed by many of today's molecular systematists, who tend to be "taxonomic splitters" because of differences they see in DNA analyses. The debate over how broadly or narrowly to delimit species in taxonomy continues to this day, however, and is still largely a matter of personal opinion.

Mayor followed *Medusae of the World* with more systematic research on jellyfishes, publishing a paper on specimens collected by others around the Philippines and Torres Straits in 1915 and another on scyphomedusae from the Philippines and the Malay Archipelago, collected by the U.S. Fisheries steamer *Albatross,* in 1917. He also conducted systematic studies on siphonophores and corals late in his career, investigating those of the West Indies during 1917–18.[40] By then, however, his main research interests had shifted away from systematics to physiology and ecology of jellyfishes and other invertebrates and to the growth of corals and development of coral reefs. Indeed, the discipline of systematics at that time was in precipitous decline, being widely viewed with disdain in academia as old-fashioned. Nevertheless, Mayor never joined in the chorus of condescension toward taxonomy and systematics. He knew that the field had an important place in science and reiterated support of it in his annual report of the Department of Marine Biology for 1915, saying, "the [Tortugas] laboratory would be remiss in its duty did it neglect the opportunity to afford facilities to competent students for carrying out work of this character."[41] Unfortunately, few availed themselves of the opportunity to do such work through the Tortugas Laboratory. In any case, Mayor's singular three-volume systematic monograph on medusae represents some of the finest work sponsored by the CIW Department of Marine Biology, and it certainly constitutes an important and enduring part of Mayor's legacy in advancing knowledge of nature.

While he was gradually turning his attention away from jellyfishes to corals, Alfred Goldsborough Mayor could rest content that his *Medusae of the World* represented a noteworthy achievement in biology. In the preface to the work he had said, "I have always felt that each working naturalist owes it as a duty to science to produce some general systematic work, and this has been an actuating motive in the production of this book."[42] Indeed, he had followed through on that precept, and his treatise on medusae became an exceptional and lasting contribution to the advancement of scientific knowledge.

V

*Remote Reefs*

With his volumes on medusae now in print, Mayor could look forward to new projects in 1911. The year began well with the publication of his sketch of the life and contributions of Alpheus Hyatt in *Popular Science Monthly.* Mayor had admired his father-in-law, who died in 1902, and he lauded Hyatt as "a fearless and independent thinker." He also stressed Hyatt's efforts to promote the theory of evolution, though he maintained that Hyatt followed Lamarck more closely than Darwin. Again, however, as Ronald Numbers has observed, contemporary views on the theory often consisted of a confusing composite, and Hyatt's notions about evolution were baffling even to those who knew him best. In any case, soon after the article was published, events took an unpleasant turn for Mayor as Hyatt's only son, Alpheus Hyatt II, died suddenly. The younger Hyatt suffered from alcoholism, an addiction that Mayor, like many others, viewed as readily curable by exercise of willpower. Although Mayor had been critical of his brother-in-law's inability to control his consumption of alcohol, he liked the man. More bad news followed when, on June 22, 1911, Harriet wrote to Alfred at Loggerhead Key that a physician had advised her to go to a resort near Bridgeport, Massachusetts, for up to six weeks for treatment of "weak lungs" and either "whooping cough" or a lingering bad cold. Two days later, however, a more astute medical doctor recommended that the illness should be "treated as a case of pulmonary tuberculosis." Harriet was advised to enter the sanatorium for tubercular patients in Sharon, Massachusetts. In a letter from Tortugas dated July 2, 1911, Alfred told Harriet that tears came to his eyes as he thought of the illness of the person who represented "all I love and long for." He wrote that he would come to be with her as soon as the laboratory's season ended on August 1.[1]

Despite the awful affliction, the ever-resilient Harriet had managed to remain active. Somehow, amid her duties as a mother of four children, including a toddler, she continued to find time to do watercolors and to sculpt. One of her works was a "model of the Neanderthal Man." In February 1912, while she was in the Sharon sanatorium, Mayor wrote from their home to inform her that the "two boxes of [your reproductions of] Neanderthal men arrived . . . [including] the original." He added, "Your restoration is based upon measurements of the fragment of the skull found . . . near Dusseldorf and the two skulls found in the cave of Spee in Belgium." Moreover, he noted, she had followed the descriptions "of some of the leading anthropologists, and the cast has already been sold to many of the leading museums of Europe." Tuberculosis had not robbed Harriet of her artistic skill, nor had it diminished her independence. Committed to improving the lot of women, Harriet championed woman suffrage and was not afraid to tell the wife of the noted conchologist William Dall that the movement for granting the franchise to women would prevail and that Mrs. Dall might as well get used to it. She often defied medical advice and took exceedingly long walks by herself or with Alfred when he visited. Painful injections of tuberculin gave her hope that the disease could be arrested for a two-year period.[2]

Earlier in the year Mayor had received a letter from William E. Ritter, who had set up temporary marine laboratories in lower California during the 1890s. In 1903, with financial support from the wealthy newspaper owner E. W. Scripps and his sister Ellen Scripps, Ritter had established the Marine Biological Association of San Diego, which was moved to La Jolla in 1905. Recipient of a PhD in zoology from Harvard just as Mayor was beginning his work there, Ritter had been at the University of California since 1893. In 1899 he was a member of the famed Harriman Expedition to Alaska. Ritter was seeking stronger support for his fledgling laboratory and thought he might receive a grant from the CIW for that purpose. Thus, he approached Mayor in 1911, asking whether he might wish to become the director of the La Jolla station and appealing to him to support his request to the CIW for a grant. While expressing esteem for Ritter and his laboratory, Mayor declined to consider the position. He offered to use his influence with CIW president Woodward to help Ritter, however. Nothing came of the effort, but the La Jolla laboratory was deeded to the University of California a year later as the Scripps Institution of Oceanography.[3] The SIO subsequently became one of the premier oceanographic research centers in the world.

In May 1911, before he had left for Tortugas and while the laboratory was being repaired, Mayor went to Jamaica to collect ctenophores, or comb jellies. Since he had decided to omit them from his *Medusae of the World,* he was planning to do a small book on those inhabiting the waters of the North American Atlantic. In Miami, Florida, soon thereafter he found the new yacht ready for service. Christened the *Anton Dohrn,* after the distinguished marine biologist

and director of the Stazione Zoologica, the 70-foot (21.3-meter) yacht was con-
structed of Madeira wood and "the best yellow pine." The Miami Yacht and
Machine Company built it at a cost of twenty-five thousand dollars. Each of its
two fifty-horsepower gasoline engines drove a separate propeller. The vessel was
13 feet (4 meters) longer than the *Physalia* and had more than double its power.
Capable of cruising at ten knots, it was "a boat of the best sort," said Mayor. The
*Anton Dohrn* proved to be a first-rate vessel for the Department of Marine
Biology of the Carnegie Institution, although Mayor decried the need and
expense of using it so often simply to shuttle people and supplies between Key
West and the Tortugas Laboratory. He clearly wished to use the *Anton Dohrn* to
expand the scope of research at the laboratory to include the developing field of
oceanography. Noted Mayor, "periodic interruption of the scientific work which
the yacht should perform has largely prevented our using this excellent little vessel for the oceanographic studies for the prosecution of which she is so well
fitted."[4]

In Tortugas by early June, Mayor set to work on his continuing studies of the
effects of starving specimens of *Cassiopea* and, in mid-July, on the annual
swarming of the palolo worm. He also supervised the spreading of another load
of New Jersey soil, which, he observed, "kills the glare and makes it [the laboratory area] much more attractive." In addition he planted more than one hundred palms and various shrubs to replace those destroyed by the hurricane of the
previous October. Among the newcomers to the Tortugas station was the young
English pathologist and bacteriologist George Harold Drew, who quickly won
favor with Mayor. Drew was interested in denitrifying activity in seawater and
the role of the marine bacterium *Bacillus calcis* (=*Pseudomonas calcis*) in the precipitation of calcium carbonate in seawater.[5]

In his report to the CIW in 1911, Mayor pointed out that additions to the
laboratory facilities had made it possible to accommodate fourteen researchers
at any time. In the same paragraph, however, he stressed the importance of
broadening the mission of the laboratory. "No geographical limits can be set
upon research," he said, "and an expedition to the Great Barrier Reef region of
Australia has become necessary in order to extend . . . the researches of Vaughan,
Harvey, Cary, Drew, and Tennent which have been commenced at Tortugas."[6]
Just why the Tortugas Laboratory was responsible for helping these scientists to
broaden their studies was not made clear by Mayor. It seemed sufficient to him
that because the men had done research work at Tortugas, they should be supported on a larger scale. Of course, Mayor's plan was good for biological research
as a whole, but it was not part of the one on which the laboratory had been
established. It cannot be overlooked, moreover, that Mayor pushed the new plan
in part for selfish reasons: to follow his own developing interests and to enhance

his role in promoting marine zoology and ecology. Personal ambitions are not necessarily detrimental to the goal of an institution, of course, and in this case they led ultimately to the elevation of work under the aegis of the Tortugas Laboratory, though they were tied so closely to Alfred Mayor that they could founder in the absence of his leadership.

During the fall of 1911, while Harriet was trying to recover, Mayor completed the manuscript of his *Ctenophores of the Atlantic Coast of North America,* which appeared in May 1912 as a publication of the Carnegie Institution of Washington. This work also became a classic, and it remains the most authoritative and comprehensive work on western Atlantic comb jellies. Consisting of fifty-eight pages, sixteen colored plates, one black-and-white plate, and twelve black-and-white text figures, the work provides information on twenty-one species of ctenophores. Mayor had personally studied eighteen of these and provided descriptions from the literature of the others. In *Ctenophores,* Mayor introduced four new taxa that are currently recognized as valid. The work also included one other species he had described in 1900. The descriptions in *Ctenophores* are thorough, and the colored drawings are outstanding. Mayor drew all of the figures and colored those in the plates. He noted that in most cases the drawings were made from living animals, which he collected along the coast from Newfoundland to Florida and on to Jamaica in the Caribbean Sea. Thanking the late Alexander Agassiz for encouraging his study of the ctenophores, Mayor expressed "respect and gratitude" for the "generous aid and unselfish interest" of his mentor. He also paid tribute to Commodore William H. Beehler, U.S.N., by naming *Tinerfe beehleri* in his honor. Commodore Beehler, former commandant at Key West, was "for many years the most constant, kindly, and powerful friend of the Tortugas Laboratory of the Carnegie Institution of Washington," according to Mayor. At last Mayor had completed the study initiated twenty years earlier, and the scientific world is the richer for his persistence. It was another superb monograph and a model of sound systematic research. Indeed, seven decades later two authorities on ctenophores noted that they found Mayor's monograph "indispensable" to their own studies, and they praised Mayor's "accurate descriptions and beautifully detailed illustrations" (see color plates).[7]

Another of Mayor's publications appeared in the same year. Although presented before the Seventh International Zoological Congress in August 1907, the proceedings of that meeting did not appear in print until 1912. Long fascinated by the annual "swarming" of the palolo worm, Mayor was now much more knowledgeable about it than he had been when he made serious mistakes on the subject in an article he published in 1900, in which he erroneously stated that the entire worm, not just its posterior section, rises to the surface during spawning time. He also confused the worm's anterior section with its posterior. In addition he had described it as a new species, *Staurocephalus gregaricus*. Mayor had corrected his mistakes in an article published in 1902 and now sought to add more information about the annual swarming of that polychaete species, *Leodice fucata* (=*Eunice fucata*), which he had first observed on June 9, 1898, and since seen on eight other occasions. He had published a third paper on the topic in 1906.[8]

In his fourth paper on the worm he reported the results of his experiments on the effects of light on the so-called swarming behavior of the Atlantic palolo, which emerges from its habitat of dead coral rubble within three days of the last quarter of the moon between June 29 and July 28. Its epitoke (the posterior end, or sexual section) appears first, after which the worm twists until the section breaks free. The epitoke, filled with eggs or sperm, depending on the sex of the worm that produces it, swims until sunrise and then bursts. Mayor collected some of the epitokes shortly after they emerged and placed them in a darkened room. He discovered that there was a delay, but not a halt, in the bursting, which led him to conclude that light was a contributing cause but clearly not the sole factor. Mayor also detailed various other palolo experiments he had conducted during the previous five years, but he indicated that he had not succeeded in discovering "the source of stimulation of the phenomenon." Although he remained fascinated by the subject, he never returned to the study of the palolo thereafter. Mayor also conducted several studies of the effects of depriving the scyphomedusa *Cassiopea xamachana* of food and light. He selected this upside-down jellyfish because it contains a "large mass of gelatinous substance" in which the algae, or zooxanthellae, are prominent and abundant.[9]

In February 1912 Mayor sailed to Jamaica aboard the *Anton Dohrn* to collect specimens and to explore the possibility of opening a branch laboratory at Montego Bay during the winter months. In May he collected specimens around the Bahama Islands. The 570-mile (917-kilometer) cruise of the *Anton Dohrn* included an opportunity for T. Wayland Vaughan to study the coral reefs off the coast of Andros Island. It also allowed Paul Bartsch, George Washington University professor and curator of mollusks in the Smithsonian Institution's U.S. National Museum, to "collect many thousands" of snails in the genus *Cerion* for the purpose of studying geographic variations. Other biologists accompanied Mayor on these collecting expeditions, and some of them worked at the Tortugas Laboratory during the season of 1912. Altogether, seventeen researchers worked under the auspices of the Carnegie Institution's marine biological laboratory in 1912. Mayor's success in attracting first-rate scientists to the facility was noteworthy. Among those who worked with Mayor in 1912 was Hubert Lyman Clark, a highly respected echinologist who had become a curator at the

MCZ in 1905. Clark had originally hesitated to accompany Mayor to Jamaica because he wanted to bring his wife along. He had asked Mayor whether it would be "a 'stag party' as . . . at Tortugas" and whether Mayor would allow Harriet to go along.[10] Of course, Clark was unaware of Harriet's illness, but once again one of the serious drawbacks of the Tortugas location had come to Mayor's attention.

Mayor had previously alluded to the importance of sponsoring long-range projects under the auspices of the Tortugas Laboratory, and by 1912 he had two men in mind as candidates for such research and for longer-term positions. One was the young Englishman G. Harold Drew, and the other was the American geologist T. Wayland Vaughan. Mayor had observed the work of both, and he liked what he saw. Holding a low-paying position in the pathological laboratory at Cambridge University, Drew might have been tempted by an offer proposed by CIW president Woodward. Because Mayor had spoken so highly of Drew, Woodward suggested that Mayor should make him a permanent member of the laboratory staff and pay him an annual salary of two thousand dollars. Despite his esteem for Drew, Mayor believed that he would not want to move to the United States because Drew not only felt an obligation to care for his parents but also held "views of life . . . hardly in accord with those of our people." Moreover, he noted, while Drew had expressed a hope that he could come to study at Tortugas every summer, Mayor believed that the young scientist was serious when he said that "his heart [is] in England." Mayor dropped the matter but continued to try to get Drew to return to Tortugas for briefer periods of research. Mayor had sympathy for Drew, but he apparently failed to grasp the depth of Drew's psychological depression when, in a letter to Mayor on October 30, 1912, Drew expressed great displeasure with his job in England and claimed that he must get out of "this damned, damp, dismal, dolefully depressing, dripping climate."[11]

Mayor had a similar plan for Vaughan, and on August 22, 1912, he sent a proposal to Woodward saying that he believed Vaughan could be lured away from the USGS if the Carnegie Institution would agree to pay him three thousand dollars per year for the next twenty years. As Mayor noted, Vaughan was doing pioneering studies of coral reefs. In fact, added Mayor, "this work is unsurpassed . . . in the world." Praising the "intensive methods" followed by Vaughan, Mayor noted in particular Vaughan's success in measuring the effects of temperature and other factors on the growth of corals. Mayor also submitted a copy of Vaughan's detailed, ten-year plan for studying the corals and coral reefs of the Caribbean Sea, the Florida Keys, and the Great Barrier Reef. Although Woodward was sympathetic to Mayor's proposition, he believed that it would be preferable to support Vaughan by grants for specific projects rather than offering him a permanent position.[12] Thus passed an opportunity for the CIW to appoint another full-time researcher of outstanding ability to the staff of the Tortugas Laboratory. Given the caustic character of Vaughan, however, the decision not to appoint him may have saved Mayor a good deal of worry and frustration.

Near the end of May 1912 Mayor received a letter from Harriet indicating that a physician had pronounced her fully recovered from tuberculosis. The news "brought tears of joy to my grim old eyes," he replied. The pronouncement proved to be premature, however, and Harriet remained at the Sharon sanatorium. For some time both she and Alfred had decided that she might recover better in Europe, and by July 1912 she was on her way, accompanied by the Mayor children and her mother.[13] It was a severe sacrifice, personally and financially, but the possibility of Harriet's recovery was something they could not ignore.

For some time Mayor had been entertaining the idea of doing research on the Great Barrier Reef. His desire emanated in part from his belief that the Tortugas Laboratory would profit from enlarging its field of investigation and in part from his increasing interest in corals and coral reef ecology. Vaughan's work had doubtless influenced him to a great degree, but Mayor was also ready to move on to the study of another group of animals in the phylum Cnidaria and to redirect his research focus from systematics to other areas of biology. Woodward liked the idea of a scientific expedition sponsored by the Carnegie Institution, and he promised Mayor that he would urge the trustees to fund it. The proposal appealed to the trustees, and by December, Mayor was making plans for the trip, to begin in July 1913. Quite naturally he invited Vaughan to accompany him, for no one was better prepared to study the stony corals off the northeastern and eastern coasts of Australia. Vaughan readily consented.[14]

The mission of the Tortugas Laboratory and its director was greatly expanded by this decision. With his monographs on medusae and ctenophores now behind him and with the prospect of an expedition to Australian waters before him, Mayor was eager to set out to sea again. First, however, he must see his beloved Harriet and the children, so he embarked for Europe in late January 1913. There he also hoped to talk with Harold Drew about the forthcoming expedition to Australia. Tragic news greeted him soon after he arrived in England, however, for he learned that Drew had died on January 29. In a letter to Harriet, by then settled in Freiburg, Germany, Mayor said, "I have been greatly shocked . . . to receive the news of the accidental death of our friend." Some people told Mayor that they believed Drew had committed suicide, for he was "found dead with a bottle of chloroform in his hand." To Mayor, however, the death of his colleague by his own hand seemed highly unlikely because he had received a letter from Drew, written on the day of his death, indicating that he was "full of plans for the future," expressing delight over the pending visit by Mayor, and speaking enthusiastically of the Australia expedition. Mayor had forgotten, or chose to ignore, the morose mood revealed by Drew in a previous letter.[15]

In any case, Drew's death not only represented a personal and professional loss to Mayor but also heightened his gloom about the health of his wife, whom he expected to see on February 19. In a letter from Freiburg, Harriet told Alfred that he should not despair, for while they had encountered severe problems, others were not as fortunate as they. After all, she said, some people have no children, others face debilitating illnesses, and still others have to struggle against poverty. Mayor soon left for Freiburg, where he spent a full month with his family. He found Harriet in better health than she had enjoyed in several years.[16] Thus, he could return to the United States with a sense of ease and resume his work with renewed enthusiasm.

As he made ready for another season at Tortugas and an expedition to the Great Barrier Reef, Alfred Mayor became concerned about the health of Wayland Vaughan, who had informed him in early January that he must back out of the planned trip to Australia. "I am on the ragged edge of nervous prostration," wrote Vaughan to Mayor, noting that he had many "partially unpublished results of investigations" from his Tortugas studies and about fifteen years of work to do in completing manuscripts of research reports for the USGS. He added that he expected to complete five monographs, totaling around twenty-five hundred typewritten pages, by the first of June 1913 and said that he was "sad over the entire predicament."[17]

Discouraged over this setback, Mayor told CIW president Robert Woodward that he might have to cancel the Australian expedition, but he quickly changed his mind. Now interested in corals himself, he believed that he could gather the information wanted by Vaughan and, in the process, launch studies of his own. He would know where to start and what to do, for Vaughan, not wishing to pass up an opportunity to expand his knowledge of corals and coral reefs, spelled out the kind of data Mayor should gather from the Great Barrier Reef. Noting that much important work on that vast reef had already been done, mostly by British scientists, Vaughan proceeded to list other things that needed to be investigated. Differences between the coral species in the Caribbean and those of the Great Barrier Reef and the greater abundance of coral species in the latter made it essential, he told Mayor, to make "a collection of Australian corals with reference to habitat." Also needed, Vaughan said, was information on the rate of growth of some of the Australian corals so that he could make comparisons with Caribbean species. Vaughan suggested that

Mayor specifically study "the post-embryonic development and general morphology" of species of corals in eight genera, and he urged Mayor to do some
"planting experiments" in order to determine the effects of water depth, amount
of sunlight, and other factors affecting the growth forms of corals. Reminding
Mayor that "certain corals exhibit a considerable plasticity," Vaughan asked him
to remove some coral clusters, cement them to concrete blocks, and "plant"
them at various levels in order to determine the influence of depth in altering
their shape. It was a technique developed by the British naturalist and, later,
noted anatomist Frederic Wood Jones and successfully used by Vaughan.[18]

After visiting with Vaughan in Washington in late January, Mayor decided
that he would not try to get him to change his mind about going to Australia.
Instead, he encouraged Vaughan to concentrate on his work at Tortugas.
Vaughan did indeed return to the station—his sixth season there. By early June,
however, his health was even worse. "Vaughan is almost a wreck . . . for he is
killing himself," wrote Mayor to Harriet. A month later, in a letter to Woodward, Mayor said, "Vaughan's health is decidedly broken." In his opinion, the
suffering scientist must take a rest, lest he "collapse mentally and physically." He
was worried, moreover, that Vaughan "may never be able to bring to a conclusion, and prepare for publication, the results of his past five years of study at
Tortugas unless we provide especially for the work." Mayor thus suggested to
Woodward that he should set one thousand dollars aside in the next budget to
be paid to the meticulous researcher on completion of the study. The publication of Vaughan's excellent work was, as Mayor noted, "of first importance—the
only conclusive study ever made upon the growth rate and ecology [of coral
reefs]." If Vaughan failed to complete the study, the Tortugas Laboratory would
"suffer the saddest loss we could sustain," Mayor wrote. Woodward agreed, and
Mayor informed his diligent associate of the proposal.[19]

Vaughan responded about two weeks after receiving the proposal, though
not in time for Mayor to reply before he departed for Australia. He expressed
gratitude to Mayor "for the friendly expressions," and then he launched into
a lengthy criticism. First, he complained that Mayor had abandoned his "corals
and the coral reef problem in a broad way" and was now stressing the biology
of them. Further, he accused Mayor of saying nothing in his last annual report
about "my uninterrupted studies of geologic process" and of failing to see the
necessity for "general oceanographic, physiographic, and geologic investigations." These charges came as a shock to Mayor, who had, of course, not only
praised Vaughan in annual reports and elsewhere but also sought special com-
pensation for him. Vaughan had not finished his lament, however. Far from
viewing the offer as a gesture of support, he construed it as an attempt to push
him more toward work on the biology of corals and into publishing his studies

before they were complete.[20] In a sense he was right about Mayor's emphasis on biology, which was, of course, the focus of the laboratory. In fact, Vaughan viewed himself first and foremost as a geologist, but he was working in an important area of overlap between the disciplines of biology and geology. Moreover, as he pointed out to Mayor, his Tortugas studies had thus far included considerable work in coral biology.

In his long missive Vaughan aptly noted that the Tortugas Laboratory had a special opportunity to conduct seminal work on both the biology and the geology of coral reefs, but he let Mayor know that his studies would continue without support from the Carnegie Institution. The grumpy geologist had missed the point, of course, and he certainly failed to appreciate Mayor's regard for his work. More than the issue of research interests lay behind the complaint, however, and Vaughan clearly revealed another reason: the meals served at Tortugas. For a while, he noted, the cook served ham and eggs, but when those were gone, the researchers, or so he believed, had to subsist on "pickled lamb's tongue," which made him sick. No one, he added, "can eat pickled lamb's tongue with impunity," and he reminded Mayor that he had "purposely omitted them from the suggestions regarding food I sent you last winter." For two days during his stay at Tortugas in 1913, he said, he had eaten "only a little rice and potato."[21]

Vaughan wanted to know why Mayor had not served more fish, but in a marginal note Mayor said, "My men complained over having too much fish" and that it had been served "twice a week" from May 28 to June 9. That amount did not suit Vaughan, and he complained that he had lost considerable weight, a point already observed by Mayor. "Before I left you," grumbled Vaughan, "I told you my principal trouble was due to starvation." Vaughan suggested that Mayor could obtain more fresh meat in Key West, which, he observed, was an item personally obtained by John Watson when he was working in Tortugas, or at the very least, he added, Mayor could "easily have plenty of live chickens brought to Tortugas." Accusing Mayor of ignoring frequent suggestions for improving the diet, he claimed that he would no longer "jeopardize my health, perhaps even my life." If he were to return to Tortugas, concluded Vaughan, Mayor must arrange for a better menu.[22]

It may be that other researchers were not happy with the Tortugas diet, but they either registered their objections indirectly as suggestions for certain foods or told themselves they could make it through a few weeks on the regimen arranged by Mayor. No one, excepting Mayor, had worked as many seasons at Tortugas as Vaughan had, though seven others had been there four or five times. The problem of serving desirable meals was in fact just one more of the difficulties of operating a marine station so far from readily accessible supplies. Although Mayor prepared menus in ten-day cycles, he might have done a better job of seeing that the variety and quality of meals suited his researchers. Lean, hardy, and accustomed to the rigors of research in isolated places, he apparently did not understand the need for giving more attention to the diet of the workers. He did recognize, however, that he might lose one of his best investigators, and soon after he got to Australia, Mayor sent a letter to Woodward to apprise him of Vaughan's complaint. He enclosed Vaughan's letter with his, as he wanted Woodward to be fully aware of the criticism. Mayor informed Woodward that he had written to Vaughan indicating that "he shall have freedom to proceed as he desires with his studies of the Florida-Bahama region, and to have such food as he desires." Vaughan "is an ill man, and the whole world looks dark to him," added Mayor.[23]

Along with fellow zoologists Edmund Newton Harvey, Hubert Lyman Clark, David H. Tennent, and John Mills, the longtime engineer and maintenance chief of the Tortugas Laboratory, Mayor had departed San Francisco for Australia on July 23, 1913. On August 15 the group reached Sydney, where they were joined by Frank A. Potts, of Cambridge University, and E. M. Grosse, an American artist who had settled in Australia nearly three decades earlier. The team was in Brisbane on August 31, and ten days later Mayor was ready to establish a base on Thursday Island in the Torres Straits, located between New Guinea and the northern tip of the Cape York Peninsula. Mayor had chosen Thursday Island mainly because he had been impressed when he first saw it in 1896 and because of information he had obtained from the "beautifully illustrated" book *The Great Barrier Reef of Australia,* originally published in 1893 by the able English marine biologist William Saville-Kent. He soon discovered, however, that currents had since swept silt and mud over the coral reefs. As a consequence, most of the corals had died, and echinoderms, the subject of Clark's inquiry, were no longer present. On September 10, before the team had fully realized the extent of change, however, Mayor wrote to Woodward, "we will work like tigers . . . , for I would much rather die than fail upon this expedition." Only four days later he penned a letter to Harriet, saying that "the reefs here prove to be too muddy for our work" and adding that the team would move approximately 120 miles (193 kilometers) east-northeast to the Murray Islands, on the seaward fringe of the Great Barrier Reef. "When Dr. Mayer determines on a course of action," Clark later wrote, "he carries it through."[24]

For the venture Mayor needed someone willing to take a boat into an area made risky by uncharted coral reefs. After much effort, he located a captain who agreed to take the job, but apparently having thought better of it, the man disappeared the next day. Not to be deterred, Mayor hired someone to take him and Mills in search of the wayward captain. They found him and his boat at Endeavour Harbour, about twenty-five miles (forty kilometers) away, and Mayor persuaded the errant boatman that he could not renege on his promise. Mayor and his fellow scientists were soon under way. The captain managed to steer the launch through reef openings and land them on Maër Island, one of the group collectively called Murray Island. The island's magistrate and schoolteacher, the Scotsman John Stewart Bruce, and the native inhabitants welcomed the party and soon turned over the local courthouse and jail for their use as a staging facility. About 3 miles long and .5 miles wide (4.8 x 0.8 kilometers) at its center, the place was ideal for research by this group of marine biologists.[25]

The island was fringed by a coral reef that, by Mayor's estimate, contained 3.6 million coral heads and showed signs of "a struggle for existence between the various kinds of coral." He also noted that the reef had grown "for ages undisturbed by severe storms." The island and its courthouse and jail served as home for the biologists for six weeks, until October 23, 1913. Soon after leaving, Mayor told Harriet, "we have lived with ex-head hunters and enjoyed it greatly," but he expressed hope that she had not heard a story "circulated in the newspapers of Australia" that the team had been murdered. He noted that both he and Clark had experienced "a signal success," while the others had "done more or less well."[26]

As he was preparing to leave Australia, Mayor sent a letter to Harriet saying that he was busy with "all the details of an intricate problem on the coral reef," and he boasted that he had "done that which Vaughan has not done, [i.e.,] made an intensive study of the reef." Although he exaggerated his claim somewhat, Mayor had, in fact, looked closely at the ecology of coral reefs, a topic to which he devoted much of his time thereafter. He sketched his studies in an article published in *Popular Science Monthly* in September 1914 and presented a synopsis of his research before a meeting of the National Academy of Sciences on February 13, 1915, but his complete study, titled "Ecology of the Murray Island Coral Reef," was delayed until 1918. The last of these was an excellent, detailed report enhanced by seventeen superb plates of sixty-five photographs taken by Mayor. Forty-eight of the photographs were of coral specimens, and seventeen were of island topography. In the article Mayor noted that Vaughan had offered "many valuable suggestions" for improving the manuscript, and he said that Vaughan's "notable work . . . [had] inspired the author at every turn."[27]

In fact, along with earlier work by Vaughan and publication in 1919 of his "Corals and the Formation of Coral Reefs," Mayor's study represents, in the words of the noted coral authority Charles Maurice Yonge, "the starting point in the modern study of living hermatypic [or reef-building] corals." To conduct his study of the Maër reefs, Mayor ran a line out nearly 1,700 feet (518 meters) from the shore across the reef, following stakes pounded into the reef floor at 200-foot (61-meter) intervals. Then, over a period of several weeks, he recorded data on the depth of the water at various tide levels, the nature of the seafloor, species of corals growing at various distances from the shore, air and water temperatures at selected times, and the effects of covering species of corals with silt. Mayor concluded that "various species of corals" were affected, in order, by (1) temperature, (2) silt, (3) mechanical effects of moving water, and (4) "the struggle for existence among the various species of corals."[28] From this pioneering study of the ecology of reefs, the determination of a correlation between temperature and coral growth was probably the most important finding. Mayor's interest in the effects of temperature on cnidarians soon resulted in other publications on that phenomenon.

In the same paper Mayor also discussed at length the theories of the formation of atolls and barrier reefs, arguing that, while it was generally correct, Darwin's hypothesis failed to explain why the bottoms of lagoons are flat and almost always at a uniform depth. In addition Mayor once again rejected the Murray-Agassiz explanation, observing that it erroneously assumed that seawater dissolves limestone faster than it actually does. He favored the theory of Canadian-born Reginald A. Daly, then a Harvard University geologist, that stressed "the control which a temporarily lowered sea-level may have exercised over the *modern* reefs," a view, he added, that was compatible with Darwin's hypothesis. Moreover, Mayor contended that Daly and Vaughan had formulated a sound hypothesis, namely that Pleistocene icecaps had lowered the water level and exposed many volcanic tips, resulting in the death of numerous coral reefs. After the melting of the ice and the attendant rise in sea level, reefs began to flourish around the submerged tips. Thus, by that explanation, present reefs had formed on top of the old, dead ones.[29]

As he had done in 1896, Mayor observed Aborigines during this visit, and desiring to compare them with a similar group, he headed for British New Guinea when he left Australia. He also had in mind writing a series of popular articles on the peoples of the South Pacific islands. From New Guinea he sailed on to Java, again to observe the peoples and culture of that land. There, on December 1, 1913, he wrote to Harriet that he would go to Colombo, Ceylon, thence to Europe, expecting to arrive in Genoa on the third day of the new year. He would then set out for Naples, where he hoped that Harriet could meet him. Due to the absence of correspondence during the next two months, it is unclear whether Harriet came to Naples or whether Alfred traveled to Freiburg. In any case, by early March, Mayor was in Cambridge, England, where he met with the coral specialist John Stanley Gardiner and presented a lecture on coral reefs. In Oxford soon thereafter, he attended a dinner in his honor as director of the Tortugas Laboratory and, in his words, "met a great many distinguished men."[30]

Back in the United States in April 1914, Mayor polished his report to the CIW trustees on the Australia expedition and on work at the Tortugas Laboratory during the previous year. Regarding the latter, Mayor noted that "conditions at Tortugas have changed in a manner adverse to the welfare of the laboratory," namely the complete abandonment of Fort Jefferson by the U.S. Navy, which made it "necessary for the *Anton Dohrn* to make at least one trip each week between Key West and Tortugas." The trips were expensive, observed Mayor, and they took the yacht away from scientific work. Thus, he recommended not merely a branch station but also "gradually abandoning the Tortugas as a site for our principal land station and establishing a laboratory upon Jamaica." Toward that end he proposed to take fifteen researchers to Jamaica in 1915 to see how it might work. Jamaica, he repeated, is "an ideal site for a truly international laboratory."[31]

Among those whom Mayor invited to Jamaica was Frederic Wood Jones, who had done pioneering work on coral transplanting, but the British naturalist and physician declined, saying that he was no longer active in coral studies. By then Mayor was preparing for the tenth season at Tortugas, with a three-week trip to the Bahamas as part of the mission. The Bahamas trip was intended primarily to let Vaughan check on the corals he had planted at Golding Cay. It would prove to be a trying experience for Mayor, for Vaughan's feathers were still ruffled. On April 18 Mayor wrote to Harriet that "Vaughan is 'cutting up' . . . again . . . , [sending me] a hot letter demanding that I explain why I invited [F. P.] Gulliver to study at Tortugas." Although Gulliver, of Washington, D.C., was interested in shorelines and had published two books on the subject, Vaughan believed that he himself was better qualified to study the geology of shorelines. Mayor had told Vaughan in 1910 and in 1912 that he hoped to get Gulliver to Tortugas, and Vaughan had said nothing, though Vaughan was now claiming that "he felt *inclined* to object." Mayor was growing weary of Vaughan's unpleasant disposition, and he told Harriet that Vaughan "is a conceited ass, and the sooner I get rid of our loathsome associate the better."[32]

Only two days later, in another letter to Harriet, Mayor complained of Vaughan's vacillation over taking the trip to the Bahamas, noting that he had already spent one thousand dollars of the laboratory budget in getting matters in order for Vaughan's research there. Mayor added that Woodward was also growing impatient with Vaughan, charging that he "thinks his is the whole Department and his work of paramount importance." The struggle continued. "Vaughan says he will in future . . . *after* this year go to Tortugas under 'other' auspices than mine," said Mayor in a letter to Harriet on April 24. Just how he would do so, said Mayor, is unknown: "Swimming, I suppose, for I know no other way." Vaughan had "quarreled with almost every one," added Mayor, "and needs a rest." In Mayor's view, "it is best for him to go in a flash of light and be done with it."[33]

Vaughan arrived in Tortugas at the end of May 1914, and with him he brought fifteen dozen eggs and a supply of codfish, though as Mayor noted, both items were on the laboratory's menu. Before he could consume many of the eggs, however, they rotted and burst. On May 7 Mayor and Vaughan set out in the *Anton Dohrn* for Nassau and Andros Island, and by the end of the month Vaughan had once again gathered important information from his research. It had become clear to both Mayor and Vaughan, however, that when the latter departed from Tortugas in June, he had no intention of returning to the laboratory. The relationship was not completely severed, however, for Vaughan returned in 1915, and the CIW, through the Marine Biology Department, continued to fund studies by him until 1923. One year later the diligent geologist moved to La Jolla, California, to become director of the Scripps Institution of Oceanography, which he helped to develop into one of the great oceanographic centers in the world. Meanwhile, Mayor worked on at the laboratory until the end of July 1914 and then headed for Woods Hole to present another of his end-of-season lectures at the MBL. He was quite uneasy, however, about Harriet and the children. Back in June he had told Harriet that he had a "presentiment of evil," perhaps concerning the possibility of hostilities in Europe, but she said that she was glad she had "not taken every one of his presentiments seriously since their marriage." Harriet assured him that all was well and tried to assuage his yearning to be with her and the children further by telling him that he would likely find another position that would not require such long periods of separation from them. In any case, Harriet said that she was planning to come home in September.[34]

Meanwhile, during the spring of 1914, Harriet and the children had moved to Normandy, but they stayed there only six weeks before returning to Freiburg. The purpose of the move is unclear, but it is likely that Harriet wanted the children to gain more exposure to French culture. In any case, she was not impressed with the French and told Alfred that she would like "to see the Germans give them another good thrashing—they need it." She changed her mind drastically in late July, however, when the likelihood of war seemed almost certain. Harriet hastily packed the family's bags, and they caught a train to Paris. As she sat in a Paris hotel on August 1, Harriet looked out a window, witnessed "a scene of wild confusion," and wrote in her journal, "the country is paralyzed with fright." Unable to book passage to the United States, she decided to stay in Rouen. On arriving in Key West on August 2, Mayor learned of the European

crisis, and unaware that his family had fled to France, he sent a telegram telling
Harriet to go to England right away. Germany declared war on France the next
day. Alfred did not hear from Harriet until more than a week later, when a letter she had written to him on August 5 arrived. "I am writing in the midst of a
confused dream, a nightmare of haste," she said. The banks were so crowded
that she could not get inside one of them to convert her German currency. Harriet told Alfred that she feared the Germans would overrun France and invade
England.[35]

Alfred continued to send letters to Harriet, but he was uncertain whether any
were reaching her. In the meantime Harriet noted in her journal that she had
"laid in provision[s] and sterilized water," packed a rucksack for each child, and
was "ready to slip on at a moment's notice." Alfred received letters from her on
August 10 and 13 urging him not to try to come to Europe and assuring him
that she and the children were safe. The German and French newspapers, she
wrote, are "inflaming their countries with fearful reports." For some reason
Mayor thought that Harriet and the children would be in England by August
14, and he sent off a letter to Frank A. Potts, who had been with him in
Australia less than a year earlier, asking him to assist his family. He added, "Our
hearts are with you all in this wicked war the Germans have forced upon you,"
and said, "I wish I could fight along with you." At last Mayor's concern over his
family abated when he received a letter that Harriet had written to him on
September 3 noting that she had gone to Havre and booked passage to the
United States on a liner scheduled to set sail on September 12. Forgetting her
earlier desire for the Germans to thrash the French, she called the Germans barbarous and wrote, "the slaughter is something never equaled in history." Alfred
had already come to share her negative opinion of Germans, but his hatred of
them intensified even more quickly than hers.[36] After the harrowing experience,
Harriet and the children got safely back to the United States—and indeed she
returned in good health.

Mayor could now concentrate on completing his report to the Carnegie
Institution trustees on the accomplishments of the Tortugas Laboratory. In
addition to comments on the Bahamas and Australia expeditions, he reported
on activities at the station in 1914. Sixteen researchers, not counting himself,
had worked at the laboratory during the season just passed. Clearly, the laboratory was holding its own. Mayor enhanced the report by providing a summary
of achievements for the ten years of the station's existence. The account noted
that forty-nine individuals, excluding Mayor, had worked there and that
twenty-eight of them, not including the director, had returned at least once. Six
volumes of *Papers from the Tortugas Laboratory of the Carnegie Institution of*

*Washington,* containing forty-seven articles, had been published, and thirty-two papers, including eleven by Mayor, had appeared as separate publications of the CIW. In addition many research reports were published in CIW yearbooks.[37]

Another forty-four papers resulting from Tortugas research had been published in other journals. Excluding papers and reports in the Carnegie Institution of Washington yearbooks, then, 104 scientific papers had resulted from work at the Tortugas Laboratory during its first decade of operation. Mayor listed the name of each investigator, the year(s) he worked at Tortugas, the topic(s) of his research, and the place(s) where he published his findings. A year later he noted that "the Laboratory has been criticized for the apparently aimless and heterogeneous character of its research work," but he did not indicate the source of criticism. Apparently sensitive to the charge, however, he justified the variety of interests by saying, "in such matters we are forced in some measure to be opportunists," and by stressing that "certain problems can be attacked at Tortugas as nowhere else in the world." Research that could be done elsewhere was not encouraged, he added. Once again, in his report for 1914, Mayor returned to the matter of location, stating that the Tortugas station must become "a branch laboratory" and a main station set up in "some such island as Jamaica." Seeking further justification for the proposed move, Mayor noted that as of 1914 the CIW had allotted less than $14,000 for structures on Loggerhead Key. Most of the $213,869 in funding had been for operations. He did not count the outlays for the *Physalia,* the *Anton Dohrn,* and other, smaller, vessels, which amounted to nearly $35,000. In his opinion, the problem of relocating the laboratory should not be viewed as enormous expenditures for buildings. Moreover, the Tortugas site was not to be abandoned; it would still be used "for certain researches."[38]

Enjoying the esteem of his peers, Mayor had been elected as vice president of the zoology section of the American Association for the Advancement of Science (AAAS) in 1913. Scientists, he told Charles Davenport in June 1914, should support the organization "our fathers bequeathed to us." To that end, he had organized a program for the AAAS meeting in 1914 designed to highlight the nature of eugenics research and the value of biology to the improvement of humankind. Mayor invited Davenport to speak on "the use of records in the study of heredity." He also asked Stewart Paton to present a paper dealing with research on mental abilities and wanted Edwin G. Conklin to give a paper on "the cultural value of the study of Biology." In addition Mayor invited George Howard Parker, a Harvard University zoologist and author of *Biology and Social Problems* (published in 1914), to address the problems of eugenics investigations.[39]

As vice president of the zoology section, Mayor was obligated to present an address, and he chose to use the opportunity to remind his fellow scientists of their social responsibilities. Urging them not to consider "the public as something colossal, crude, and struggling, something far and apart from our cloistered world," he said that scientists should maintain an attitude of humility, remembering that they are "simple men and women" who could profit from the assistance and advice of the public. Getting to the real reason for his plea, Mayor added, "We now stand as trustees guarding things of vast import for good or evil . . . , [for] the very word eugenics conjures up problems for the wisdom of humanity to solve." Militarism was not confined to Germany, he said, noting that it was also present in the United States. Scientists must therefore remember the importance of "mutual helpfulness between our public and our men of science."[40] Despite the racial biases and the social-Darwinist views he shared with many of his contemporaries, Mayor was at heart a humanist, even if a paternalistic one.

Meanwhile, Mayor's hatred of Germany had grown even stronger, fueled in part by a letter from John Stanley Gardiner, who spoke of the war as "clearly a life and death struggle" for Britain. In Gardiner's view, the worst thing the Germans had done was to invade Belgium, resulting in "the rape of the aged and immature." Gardiner suspected that reports of German barbarity "rather tears you [Mayor] in pieces, as you . . . were fond of Germany." Indeed, Mayor was becoming increasingly touchy about his Teutonic surname, and he eventually came to believe that it made him suspicious in the eyes of some people. He was perhaps more sensitive to the matter than was warranted, but as he prepared early in 1915 to do research in the West Indies, he urged Woodward to write a letter for him stating that although his name was "of German origin, . . . he and his forbears for several generations have been American born and American citizens." News of the German sinking of the *Lusitania* on May 7, 1915, further aggravated Mayor's sensitivities about having "a Hun name."[41]

Back in March 1915 Mayor had booked passage to the Caribbean, and throughout the month of April he examined coral reefs around the Virgin Islands, Antigua, Dominica, and other islands and considered the suitability of each place as a site for the Carnegie Laboratory. In May he piloted the *Anton Dohrn* to Puerto Rico to investigate that island as a potential site and to make it possible for six biologists to conduct research there. Finding fewer coral reefs and much mud and silt surrounding the living ones, Mayor ranked Puerto Rico far behind Jamaica as a desirable place for the marine station. He also rejected it for another reason, though it was a matter he revealed only to Harriet. Apparently hoping to be able to have Harriet and the children live with him if he succeeded in getting the CIW trustees to move the laboratory, Mayor also considered the culture and peoples of the area. Puerto Rico would not do, he said, for it is "the worst of the tropics . . . [and] no children can be successfully raised here." Once more indicating his prejudice, he referred to Puerto Ricans as "a black and tan mixture of Negro, Indian, and Spanish."[42]

Fortunately, from Mayor's point of view, the Tortugas Laboratory had enjoyed another season of success, and in his report for 1915 he listed twenty-one researchers who had conducted inquiries either on the trip to Puerto Rico or at Loggerhead Key or both. In addition to himself, they included Edwin G. Conklin, Ulric Dahlgren, Edmund Newton Harvey, Charles F. Silvester, Gilbert van Ingen, and Alexander H. Phillips—all of Princeton University. Also present were Paul Bartsch, of George Washington University and the U.S. National Museum, and old standbys Vaughan, Treadwell, and Longley, all of whom had conducted important research at the laboratory in 1915. Mayor also noted that nine of the sixty-seven papers presented before the last meeting of the American Society of Zoologists were based on studies completed under the auspices of the Tortugas Marine Biological Laboratory. In addition Mayor had been successful in publishing during the year. Finally appearing in print, in a volume of *Publications of the Carnegie Institution of Washington,* though he had completed it long before, was his article on "Medusae of the Philippines and of Torres Straits." He later republished the article in slightly altered form in the *United States National Museum Bulletin.* His descriptions included thirty-one species collected by others in Philippine waters in 1909 and sixteen he had personally taken during the Australia expedition in 1913. Written to appeal to a broad audience, the engaging descriptions and general comments filled forty-five pages. The article included seven black-and-white illustrations and three colored plates, all done by Mayor. In fact, Mayor had described and figured some of the Philippine species in his *Medusae of the World,* but here he established four new taxa, of which three are presently valid, namely the rhizostome scyphozoans *Cotylorhiza pacifica, Phyllorhiza luzoni,* and *Catostylus townsendi.* The last was "named in honor of the author's friend, Dr. Charles H. Townsend, the distinguished director of the New York Aquarium."[43]

Once again Mayor had demonstrated his skill as a systematist. It is interesting to note, however, that in his annual report to the CIW in 1915, he said that taxonomic studies had been "relegated to a secondary place" by the laboratory, maintaining that much of that kind of work on tropical species had already been done. "The major efforts of the Laboratory," he averred, "have been directed toward experimental work in biology and physiology." Yet, in 1916 he inserted in his report to the CIW a statement to the contrary. "It is perhaps unfortunate," he stated, "that the absorbing interest which the experimental side of biology has in recent years attracted to itself should have cut down activity in systematic work." Referring to the "beautiful and valuable papers" published by the Stazione Zoologica, he asserted that the laboratory must not "neglect the opportunity" to attract excellent systematists to work at Tortugas.[44] In fact, however, the nature of the work there remained unchanged. Despite his own superior systematic work, Mayor had set a course in line with the general trend in biology. Experimentation and laboratory work had become dominant in the field of biology, and interest in systematics had continued to slide down the scale of significance. Perhaps his brief return to systematics had made Mayor aware of the overall neglect of systematists and renewed his appreciation for this vital area of biology, but it did nothing to reverse the general course of emphasis.

Building on the studies of jellyfish physiology he had begun in 1905, Mayor completed three more papers in 1915 on nerve conduction and pulsation in medusae, and he published another six articles on that topic between 1916 and 1920. This line of his research had become progressively more sophisticated and reflected his abilities in mathematics and engineering as well as in biology. Mayor carried out complex experiments to shed light on the physiology and chemistry of processes involved in the triggering of nervous impulses in jellyfishes, in the transmission of them across jellyfish nerve nets, and in the factors that influenced them. A driving force behind this work, at least initially, was likely to have been the hope of some practical application to human health and especially to cardiology, as noted earlier.

The scientific importance of Mayor's pioneering research on the nervous system of jellyfishes has been highlighted by physiologists such as L. M. Passano of the University of Wisconsin (see p. 55). Of particular importance was the Mayor discovery of the "entrapped wave preparation" (EW), or "the entrapped circuit wave."[45] Mayor subsequently employed this preparation, using the abundant jellyfish *Cassiopea* as the experimental animal, as a model system in studies on the electrophysiology of nerves and nervous systems. While this model offered potential for investigations of neurophysiology, new and improved methodologies were soon developed. Preparations using nerves from other organisms also offered fewer problems than those from jellyfishes. Mayor's EW preparation was thus abandoned. Nevertheless, Passano considers Mayor's experiments on the jellyfish nervous system to have been extensive and accurate, especially given the isolation of the Tortugas Laboratory and the primitive facilities available at the time. Mayor's accomplishments were therefore noteworthy and influential.[46] Among early pioneering investigators on jellyfish pulsation and pacemakers, Mayor deserves mention along with the esteemed British naturalist George John Romanes, who conducted a classic series of studies on the jellyfish nervous system during the 1870s.[47]

The fourth paper presented by Mayor in 1915 was an insightful approach to testing the Murray-Agassiz hypothesis that dissolving limestone accounts for the widening and deepening of atoll lagoons. Using mollusk shells, the experiment consisted of four parts: one in a sealed jar of twice-filtered seawater; another in a jar filled with nonfiltered seawater; a third in an open jar inside "a wooden dark chamber" and placed in the moat at Fort Jefferson; and a fourth in an open jar and placed slightly below the surface near the wharf at Loggerhead Key. Mayor weighed each of the shells before placing it in its respective jar and again a year later. This allowed him to calculate the amount of dissolution of the shells. He concluded that it would take nineteen million years to dissolve a layer of calcium carbonate at twenty fathoms (thirty-seven meters), the mean depth of lagoons. Admitting that sea cucumbers, echinoderms, marine worms, algae, and rainwater no doubt played roles in dissolving limestone, Mayor nevertheless offered evidence that ran contrary to the Murray-Agassiz hypothesis. His interest in the diet of holothurians later led him to additional comments on the subject and, according to the University of Southern California biologist Gerald J. Bakus, made him the first biologist to study the topic, though Mayor probably overestimated the amount of sediment dissolved by sea cucumbers.[48]

Optimistic over the work he was lining up for the Tortugas station in 1916 and, like many of his fellow Americans, thinking that the United States would not enter the war that was killing and maiming thousands upon thousands of young men in Europe, Mayor planned an expedition to Tobago, Trinidad, and Grenada to examine the coral reefs of those islands and to assess each place as a site for relocating the laboratory. The expedition ran from mid-March to April 23, 1916. Accompanying Mayor were Hubert Lyman Clark and the University of Copenhagen zoologist Theodore Mortensen, who specialized in the study of echinoderms and crinoids. Mayor used his time to collect and study corals and siphonophores. Earlier, in January 1916, he had written to Woodward to remind him to apprise British authorities of the exact nature of the expedition, for he feared that island officials might be suspicious of foreigners working around coral reefs. Three weeks later he asked Woodward to stress to British officials that he and Clark "are about as pro-British as the English themselves" and that, as "a true Dane," Mortensen "detests Germany."[49] Mayor did not want to run any risk of being mistaken for a spy.

By the end of the expedition Clark had collected specimens of seventy-three species of echinoderms, Mortensen had made some discoveries related to the embryology of crinoids, and Mayor had taken "some new stages of Siphonophores." On May 24 Mayor left for Tortugas, where seven biologists would join him for research at various times during the season. The laboratory also sponsored research by Aaron L. Treadwell, of Vassar College, who studied polychaete worms in Bermuda waters. Among the researchers at Tortugas was William H. Longley, working for the sixth time at Loggerhead Key. For his work on coloration in reef fishes, Longley had requested an underwater camera and a diving hood, the latter of which intrigued Mayor as a way of enhancing his own investigations of coral reefs. Longley also needed a twenty-six-foot launch fitted with a glass bottom for his studies. Such items were expensive, but the Carnegie Institution trustees continued to authorize the purchase of everything Mayor requested, thanks to the unwavering support of President Woodward. The annual budget for the laboratory had increased from $15,700 in 1905 to $19,000 in 1916, but this increase did not include many special grants to individuals working under the auspices of the laboratory, nor did it include additions and replacements of equipment. Buildings, seagoing vessels, and costs of maintaining and repairing them continued to run high, already amounting to $50,448. A storm on July 4–6, 1916, did no damage to the laboratory, but another one on July 26 of the same year resulted in the loss of a dory.[50] The laboratory and its equipment clearly remained vulnerable to the fickle fury of nature. Still, of the allocations to the ten research departments of the CIW, that of the Tortugas Laboratory remained the lowest by far. Although the modest annual allotments to the marine biological department were due in part to the lack of expensive buildings and a permanent staff other than Mayor and Mills, they also represented Mayor's ability to accomplish much at relatively small cost to the institution.

Meanwhile, in 1915 and 1916 Mayor was publishing articles on the peoples and cultures of Oceania and Australasia. Appearing in *Popular Science Monthly* and, later, *Scientific Monthly,* the articles and photographs taken by Mayor appealed to readers with a romantic bent toward the South Sea Islands, and taken as a whole, along with a nearly completed book manuscript later written by Mayor, they comprise a coherent, if often stereotypical, view of the region's peoples. Among those responding to the articles was William MacGregor, a British physician, colonial administrator in Tahiti for thirteen years, and from 1888 to 1898 lieutenant governor of British New Guinea. MacGregor noted that Mayor had made a few mistakes in his four-part history of Fiji but said that the history "will be referred to as standard authority." MacGregor also opined that the work was "free from prejudice and comes from a broad and open mind." While they were informative and revealed Mayor's love of the area, the articles and the book manuscript in fact reflected contemporary biases and the impact of eugenicist views. Mayor would continue to offer his observations on such matters throughout the remaining years of his life. Along with his continuing scientific studies, the papers on Oceania and Australasia enhanced Mayor's reputation, and in April 1916 the accomplished marine biologist was elected to membership in the prestigious National Academy of Sciences (NAS), before whose meetings he had already presented several papers.[51]

Mayor had Davenport to thank for nominating him to membership in the NAS. The honor prompted Mayor to propose to Davenport on October 28, 1916, that the latter "draw up a set of suggestions . . . [for] preparing biographical memoirs of members of the NAS with a view to making such memoirs of greater service as data for a study of hereditary traits." Quite naturally, Davenport liked the idea, and he soon sent a letter to each NAS member requesting family and personal information. The request was designed, without saying so, to play up the role of intellectual elites in society. The number of NAS members who responded is unknown, but Mayor was one of those who did. By then, despite advances in bacteriology, Mayor had come to believe that tuberculosis was an inherited trait. He was puzzled, however, for he could not reconcile the presence of a defective trait in Harriet along with her superior intelligence and artistic ability. Mayor also believed that he could contribute to a better understanding of heredity by writing a sketch of the life of William Stimpson for the *National Academy of Sciences Biographical Memoirs.* The prodigious marine zoologist Stimpson, who died of tuberculosis in 1872, had been "an old and intimate friend" of Mayor's father. Mayor wrote the biographical account, which was published in 1918. Stimpson, who died at age forty, was the first curator of invertebrates at the Smithsonian Institution. In his relatively brief but brilliant career he described around 950 species, one-half of them crustaceans. According to a later authority on Stimpson, "the biographical sketch [by] Mayer is basically accurate" and represents "a fine job [of] capturing Stimpson's personality."[52]

In the meantime, Mayor was planning another expedition to the South Pacific, this time to American Samoa. Once again Woodward paved the way, persuading the CIW trustees that the mission of the Tortugas Laboratory must go far beyond the waters of the Gulf of Mexico and the Caribbean Sea, and that Mayor and his associates would make important contributions to marine biology. Mayor was somewhat concerned about leading the expedition, however, for he was suffering from a persistent cough that he feared was a sign of tuberculosis, though he and Harriet had made themselves believe for several months that it was merely a bad cold. Moreover, the ardent patriot Mayor, now increasingly aware of the strong possibility of American involvement in the Great War, wondered whether he should forgo the expedition in order to serve his country in the event of war. He was willing, he told Davenport, to serve "as a private" in the American army, but he thought it more likely that he could help by commanding the *Anton Dohrn* "as a naval patroll [*sic*] boat in the auxiliary naval service, which might be of use in looking for submarines." His hatred of Germany had grown to the point that he was contemplating even more seriously a new spelling of his surname. To Davenport he wrote: "I do feel ashamed of having even only ¹⁄₁₆ of the Hun blood," but he said that Harriet was arguing that "it is better to keep the horrid name [of Mayer] and show we are really thorough Americans."[53]

Mayor hesitated about leaving Harriet and the children in time of crisis but could not bring himself to cancel the expedition. The lure of the sea, the excitement of continuing his work on coral reefs, especially underwater in a diving hood, and the thought of adding to his studies of the peoples and cultures of Oceania were powerful attractions. Thus, in February 1917 he entrained for San Francisco with his Princeton colleague Lewis R. Cary, an authority on octocorals who had gone to Tortugas several times. They soon boarded a ship for the South Pacific. As usual, the journey included a stop in the port of Honolulu, where Mayor noted that automobiles were now abundant. "Honolulu," he wrote to Harriet, "has become a big, ugly city," though he viewed the rest of the land as "a veritable paradise." On March 4 Mayor and Cary arrived in Pago Pago, and soon thereafter they began their studies of the coral reefs around Tutuila, the main island of American Samoa. They worked there until April 18.[54]

While Cary turned his attention mainly to the alcyonarians, or soft corals, Mayor concentrated on further testing of the Murray-Agassiz hypothesis of reef formation. During the previous year he had been challenged by the Harvard University geographer William Morris Davis concerning his argument that his Murray Island studies revealed flaws in Davis's hypothesis about the origin and nature of coral-reef formation. Contentious and self-assured, but bright and insightful, Davis objected especially to Mayor's statement to him in a letter written in late January 1916. "I am inclined to think," Mayor had said, "that the *final* theory will embody much that you advocate, *together* with Daly's interpretation." In an acerbic response Davis replied, "It would be an aid in my work if you were willing to specify the ground of your inclination. Coral reef literature is full of unspecified inclinations." Such inclinations, added Davis, are "subjective . . . [and] don't cut much objective ice." Ending his letter by asking Mayor whether he had "any direct evidence" for his argument, Davis pressed Mayor to offer a more convincing case.[55]

Recognizing that he needed stronger affirmation of his view, Mayor hoped to study the reefs more closely by going underwater in a diving hood, but for reasons unknown, he apparently did not do so on this occasion, although he would venture into the deeper waters at Tortugas only a few months later. He could approach the problem in part, however, by studying whether rainfall draining from a volcanic, forested island such as Tutuila would become acidic enough to dissolve reefs near the shore. He tested the alkalinity of the streams flowing from the island, trying to determine its effect on corals by placing twenty-six specimens of twelve species of corals approximately 150 feet (45.7 meters) from the mouth of a brook at Pago Pago, where no corals grow naturally. All but two specimens of a species of *Porites* were dead within fifteen days, but Mayor concluded that it was not only the freshwater of the streams but also a covering of silt that killed the corals. In addition his study showed that "rain falling directly into the sea has far more effect in reducing the alkalinity of the surface water than has stream water pouring outward from the shore." More study of the phenomenon was needed, observed Mayor, and he stated his intention to pursue his research on a return trip to Samoa in 1918. In any case, Mayor had offered further evidence against the Murray-Agassiz hypothesis, though he had not convinced Davis that he should modify his argument.[56]

As he had done on the way to Samoa, Mayor tested "the hydrogen-ion concentration of the surface water" of the Pacific each day on the return trip. He seems to have done so at the request of the University of Minnesota biologist Jesse F. McClendon, who had recently completed similar studies on the California coast. Mayor published a brief paper on his findings in the *Proceedings of the National Academy of Sciences* soon after his return. Back in Princeton by late April 1917, Mayor began to prepare for the season at Tortugas. With his country now at war, however, he planned to make the season shorter and to limit the number of researchers to seven scientists who had previously been to Tortugas. The laboratory also sponsored a research trip to Japan for Edmund Newton Harvey and gave a small grant to the Princeton University zoologist Ulric Dahlgren for the study of electricity production in some species of fishes.[57]

# Twilight at Tortugas

Despite concern over the war, unhappiness over his German name, and the cough that would not go away, Mayor kept himself busier than ever. Indeed, he taxed his physical strength more than usual. To Harriet he wrote on June 16, 1917, "I go under water in the diving hood to study corals and am carrying out some work on the nerves [of *Cassiopea*] that may lead to something." He added, however, that "somehow this year seems to be an off one for research, and no one enters wholly into the spirit of it as they once did." Nevertheless, Mayor completed and published important studies of the effects of oxygen deprivation, heat, and light reduction on corals; the effects of explosive shock on the nervous system of invertebrates and fishes; and the limestone-dissolving capacity of holothurians (sea cucumbers). His study of the impact of explosion on marine animals appears to have been for the purpose of determining the influence of "war shock" on troops. Mayor concluded that "war shock is predominantly a psychic phenomenon, and being hysteria, it can be cured by hypnotic suggestion."[1] The questionable idea reflected not only his lack of experience in combat but also a rather simplistic approach to healing a human trauma. In addition it indicated a view that a biologist could readily transfer results from studies in his field to the field of psychiatry.

On July 30, 1917, after a truncated season of two months, Mayor closed the laboratory for the season. He would not head directly for Princeton, however. Continuing to believe that he could be of service to his country, Mayor hoped that he might be "called out into active service . . . to cut through the submarine [sic] zone on a merchant vessel." The chances that a nearly fifty-year-old man with no military training and in a state of precarious health would be called into such service were not promising, but the devoted patriot hoped otherwise. In any case, Mayor had already decided to take an examination for navigators, and he headed to Jacksonville, Florida, for that purpose and to turn the *Anton*

*Dohrn* over to the U.S. Navy for use as a shore-patrol vessel to guard the harbor at Key West, a measure readily approved by the CIW trustees. Mayor easily passed the navigator's examination of one hundred questions, and on August 7 he wrote to Harriet that he had received a diploma qualifying him as a "first class Yacht captain." His knowledge would soon be put to use, for during the fall Princeton University assigned him to teach navigation to students who were contemplating commission in the U.S. Navy. Fifty-three young men enrolled for the first class.[2]

By the first of December, however, Mayor was seriously ill. A week later Harriet informed Woodward that her husband was doing better after he had experienced a "crisis and been a very ill man." She added, "He looks terrably [*sic*] thin and wan, but [he is] happy to be in this world." On December 27 Mayor was able to pen a letter to Woodward, and he recommended that they wait awhile before making any decisions about the Tortugas Laboratory. He added, however, that "it might be desirable to develop the Tortugas into the station we have in mind," though he offered no details about the plan.[3] His illness and the devastating war dictated caution.

Meanwhile, in late December, the Mayors learned of the death of Stuart Walcott, son of Charles D. Walcott, a noted paleontologist famous for his work on the Burgess Shale in British Columbia and, at the time, secretary of the Smithsonian Institution. As a longtime trustee of the CIW, Walcott was a friend of the Mayors. Walcott reported that his son Stuart, an aviator, had been on patrol and destroyed "a German two-seater" but was subsequently "overtaken by four single-seater Albatross German aircraft." According to the report given to his father, "his machine took a nose dive within the German lines." Mayor wrote to express his and Harriet's sympathy. "Bad as it all is," he said, "it must be no small consolation to have given a brave and worthy soldier to this greatest of causes in defense of national honor." A few weeks later Mayor again wrote to Walcott, praising his "heroic son . . . and his fine character, courage, and patriotism." He added that he envied America's fighting men. Soon thereafter Harriet painted a portrait of Stuart Walcott from a photograph and presented it to his parents.[4]

Alfred Mayor would indeed have taken up arms had he been suited and fit for soldiering. Unable to do so, he focused his attention on his scientific work. As he lay recovering from the illness that had threatened his life, Mayor suppressed thoughts of his mortality and began to plan for a return to American Samoa. On January 15, 1918, he requested authorization to return to Tutuila for further study of coral reefs. He wished, he told Woodward, to continue his investigation of the growth of corals and "to drill through the reef (200 feet [61 meters] deep) to determine [its] nature." Only three days later he was able to thank his faithful ally not only for granting permission for another expedition to Samoa but also for authorizing a return to Tobago. Reporting that the symptoms of his "cold" had diminished, he set off on a commercial steamer for the West Indies, arriving in Port-of-Spain, Trinidad, in mid-March. By the twenty-second day of that month he was hard at work. He planned to return to the United States by May 1, but an agent of the Trinidad Line made a mistake in booking his return and thus managed to delay his departure for eleven days, though Mayor suspected that the error was deliberate.[5]

Increasingly fearful that his studies in the region might draw the suspicion of local authorities, Mayor had urged Woodward to assure them that he had "no Hun tendencies in me despite the fact that I am cursed with 1/16 German blood." He also informed Woodward, "I have come to loathe the sight of my unfortunate German name to such a degree that I intend to take legal steps to have it changed [from Mayer] to Mayor, thus reviving a form of the old Swiss spelling." Not long after reaching his home in Princeton, New Jersey, in May, someone, perhaps Woodward, informed him that the Trinidad Line agent was apparently circulating a rumor of suspicious activities by Mayor while he was working at Tobago. In an effort to allay the rumor, Mayor swore out an affidavit on May 27 asserting that he was a loyal American citizen and that he was the instructor of a naval unit at Princeton University. He stated that he and his fellow workers had obeyed every law and devoted their time at Tobago entirely to scientific research. In addition he noted that he had nearly completed a book on navigation designed to assist young men in passing the examination required for a commission in the U.S. Navy. Neither he nor any member of his group had engaged in any subversive activity or communicated in any way with enemies of his country or Great Britain, he asserted. Mayor also asked Woodward to write a letter of support to the U.S. secretary of state, which Woodward promptly did, saying that he had known Mayor for twenty-five years, had been an intimate friend of his father, and had no doubts about Mayor's loyalty.[6]

Two days after swearing out the affidavit, Mayor wrote another letter to Woodward telling him that he had asked Trinidad's chief of police to investigate the charges against him and to send a copy of his findings to Woodward, the U.S. Department of State, and the governor of Trinidad. Suspecting that the rumor had originated with the Trinidad Line agent, he wanted everyone to know that the agent was "basing his opinion of me upon a five minutes, rather heated conversation respecting the mistake made in . . . placing us in the wrong steamer." The agent had never been "within 50 miles [80 kilometers] of the scene of our laboratory in Tobago," Mayor added. Revealing his racial bias, he said that the agent was one of the "ignorant mulattoes . . . [who are] naturally suspicious of every foreigner of white blood." The request brought the desired result, for on July 25, 1918, the inspector general in Trinidad wrote to Mayor saying that "there has never been any suspicion about you or any of your party" and expressing regret over the complaint. Not until nearly three months later, however, did the commissioner-warden and magistrate of Tobago write to say that his investigation had exonerated Mayor of any suspicion. Although he declared that Mayor and his associates "were engaged wholly and solely on biological research work from the first to the last day of their sojourn in this island," he stated that he could not believe any discourtesy had been shown toward them.[7] Mayor was at least satisfied that the record had been set straight, although in due course he would find that it had not.

By June, Mayor had completed the little manual on navigation, most of which he had written "at sea," that is, on the voyage to and from Tobago. Published by J. B. Lippincott and Company, the 206-page manual was titled *Navigation, Illustrated by Diagrams.* Intended for "young men of limited mathematical training who hope to qualify as Ensigns in the United States Navy," the manual was replete with diagrams, all drawn by Mayor. As he said in the preface to the work, Mayor aimed "to use simple language and to make it very practical." Consisting of twelve chapters, the pocket-size book dealt with such topics as use of the compass and the sextant, calculating latitude and longitude, and maritime signals. Mayor viewed the manual and his role as a navigation instructor as contributions to the American war effort, and thus they were important to the unswerving patriot. By the time the work appeared in print, the spelling of his surname had been officially changed from Mayer to Mayor, an action completed in court on August 5, 1918. It was a move that "seemed fitting," Mayor said, "to repudiate the last link of association which bound us to a nation that had forced our country into the line of its enemies." Alfred Mayor proudly noted the alteration in spelling to those who might not be aware, as he did on November 2, 1918, in a letter to H. S. Jennings: "I have changed my name so as to try to forget my $\frac{1}{16}$ Hun blood."[8]

His tubercular infection was now in a stage of temporary arrest, leading him to believe that he might be cured. In fact, Mayor became quite optimistic. After all, he might be as fortunate as Harriet, who had experienced no symptoms of recurrence. Thus, he departed in high spirits for San Francisco on June 18, 1918, and for Pago Pago soon thereafter. On July 23 he wrote to Harriet from Pago Pago that "all of our old native friends are here in good form," but he noted that many of them "have become chiefs and cannot do any more work for white men." He deemed his new native workers as "the laziest I ever saw." Reflecting on the "charming little Polynesian town" he first saw in 1898, he now witnessed "ugly . . . white men's houses with corrugated iron roofs." Still, to him, no place in the South Pacific was "more beautiful than this Harbor of Pago Pago, deep blue and shaded by high volcanic walls all green above it."[9] He would write more about the impact of Western culture on Polynesia later, but for now he must move on to his scientific work.

After a ten-day trip to study the coral reefs around Fiji, Mayor returned to Pago Pago to resume the research he had begun a year earlier. Using grids of 24 square feet (22.3 square meters) at intervals of 50 to 100 feet (15–30 meters), he calculated the number of corals on the Aua Reef. By weighing samples of living corals and then, after killing them and dissolving the soft material, weighing the skeleton, Mayor was also able to calculate the amount of limestone added each year to the reef. Aware that breakers removed some of the material, he set six open barrels near the seaward edge of the reef in order to calculate the amount of limestone carried away. He also had to calculate how much limestone was added by boring creatures and sea cucumbers, which, in his judgment, exceeded the loss by currents and thus caused the sediments of the reef flats to deepen. As they deepened, he concluded, the impact of currents lessened and more sand and mud were added to the flat. The result was that fringing reefs moved seaward, but Mayor did not believe that the inevitable consequence was the development of a barrier reef. That was a question requiring further study, especially by observing the reefs underwater in a diving hood. He postponed that step for another year, however. In any case, this investigation and the earlier one at Maër Island in 1913 constituted pioneering research in the ecology of coral reefs. Fifty years later two scientists located Mayor's transects on the Samoan reefs and completed a new survey of the same reefs. Mayor's pioneering work thus provided not only an opportunity to complete a study of the human impact on the reefs after five decades but also, as the scientists noted, the potential for even longer-range analyses.[10]

Once again Lewis Cary and John Mills accompanied Mayor—Cary for the purpose of studying the role of alcyonarians (sea whips and sea fans) in building reefs, and Mills for the purpose of drilling into the reef in order to determine the thickness of the reef and the nature of the base on which it had formed. In Pago Pago harbor, about 575 feet (175 meters) from shore, Mills was able to run the drill to a depth of 121 feet (37 meters) before hitting volcanic rock. Mills was unable to obtain solid core samples, however, as the material fragmented too easily. Mayor and his associates left Samoa on August 5, 1918.[11] With the *Anton Dohrn* in the service of the U.S. Navy and with Mayor in the South Pacific, no work was done at the Tortugas station in 1918.

After the war ground to a halt early in November 1918, Mayor was eager to get to Tortugas as quickly as possible, and he went to Key West in late December. The U.S. Navy returned the *Anton Dohrn* to Miami, Florida, on January 2, 1919, and Mayor set out immediately from there for Tortugas, where he

remained until the 16th. Checking on the corals he had planted two years earlier, he found them in good condition. At last Mayor could plan for resumption
of a regular, if abbreviated, season at Loggerhead Key, from early May until June
20, and soon thereafter head once again to Samoa and Fiji.[12] The Tortugas station had clearly become secondary to Mayor's main places of research in the
Pacific, but considerable time had elapsed since he had raised the issue of relocating it.

Early in February, in anticipation of his trip to Samoa and Fiji during the
coming summer, Mayor applied for renewal of his passport, but he learned that
the U.S. Department of State was holding it. Word came from his friend Paul
Bartsch that the American consul in Port-of-Spain had sent a telegram to the
Department of State accusing Mayor and the men who had been with him in
Trinidad and Tobago of being "suspicious characters." In a letter to Mayor, however, the consul said nothing of that matter but complained that Mayor had not
called on him. In fact, Mayor told Woodward, he and Treadwell had indeed visited the office of the consul and left their calling cards when told that the consul was away. The consul had neither bothered to "consult the local authorities"
nor even come close to the place where they were conducting their research. It
appears, then, that the trouble had not originated with the unfairly maligned
Trinidad Line agent but with a hypersensitive American official, who continued
his vendetta long after the war had ended. Mayor was all the more outraged
over the hint that he and Treadwell were disloyal. Indeed, to him, the suggestion was tantamount to accusing them of "a crime worse than murder, that of
treason to our country."[13] Eventually the passports were returned, and Mayor
felt that he could at last lay to rest his fear that anyone would ever again question his patriotism.

With Woodward once more smoothing the way, Mayor, Lewis Cary, Reginald Daly, and John Mills prepared to return to Samoa and Fiji after the season
at Tortugas. Mayor was concerned, however, that the specified allocation might
be insufficient to fund his proposed trip around the world to test the alkalinity
of seawater when the research in Samoa was done. In the meantime, at a meeting of the NAS, Mayor presented another paper on the topic. No doubt this further impressed Woodward, who replied, "If you need a few thousand dollars
additional to carry out some contemplated work, . . . I think the Executive
Committee will be willing to grant them." Mayor expressed a preference for
staying within the budgeted amount, however. He thanked Woodward for his
offer and said that his generous effort "made it a delight to serve under you." In
fact, Woodward was already laying the groundwork for the more generous
budget Mayor desired for the laboratory in 1920. What Mayor had in mind was
"to get up a better volume on coral reefs than any that has yet appeared." He

added, "with the men I have in mind I am sure we can do it." The details of Mayor's proposed project do not appear to be extant, and it is likely that he never worked them out. In any case, the broad plan was grand, important, and costly. So excited was Mayor over the proposal that he told Woodward, "I feel quite like a school boy of 16 over the prospect of 1920 and the coral reefs."[14]

The brief season for the operation of the Tortugas Laboratory in 1919 gave Mayor a chance to bring his thirteen-year-old son Brantz along. Brantz later recalled that he accompanied his father on other trips to Loggerhead Key. In 1915 Mayor had taken his older son, Hyatt, with him to the Tortugas. Both boys had been well received by the researchers. Hyatt had mounted butterflies for one of them and collected on the coral reefs, while Brantz eagerly accompanied various biologists on their collecting trips and maintained the boat motors. Mayor was, of course, proud of Hyatt, who by that time had won "Honors in English and Mathematics" as a Princeton University student, but he thought that Brantz was more likely to follow in his footsteps. Hyatt eventually received a Rhodes Scholarship and turned his attention to art. Brantz, on the other hand, followed a different career. Also a graduate of Princeton University, he worked in the aircraft industry for many years and then in the magazine-publishing business. Both daughters, Katherine and Barbara, were equally talented and won recognition for their achievements. Mayor was a devoted father, and during their early years, when the children were gathered around the breakfast table, he would often sketch an animal and describe it for them. He also read to them and related stories of his adventures. During his many excursions, his children often wrote affectionate letters to him, and when he was home, recalled his son Brantz in later years, he sat "in an armchair with us kids swarming all over him." Brantz also recalled with great delight how his father would walk with him on many evenings on Loggerhead Key to the lighthouse and there talk with him about life as they watched the sun set.[15]

Arriving in Samoa on July 23, 1919, Mayor, Cary, Mills, and Daly planned to go on to Fiji right away, but an influenza quarantine prevented them from entering the place. In fact, the epidemic continued throughout the nine weeks they were in the region. Thus, they concentrated their work on the reefs of American Samoa, surveying all of them except remote Rose Atoll. On August 11 Mayor wrote to Harriet that "Daly is working day and night over the problem of the old volcano and finds that it is more ancient than the Hawaiian Islands and greatly eroded." He also reported that he had been "very hard at work . . . , planting corals on the reefs, [at a depth of] 7–8 fathoms [13–15 meters]." Meanwhile, Mills was boring through a reef and had reached a depth of forty feet (twelve meters). Mayor was even more thrilled to tell Woodward in a letter written on September 28 that, "using diving apparatus we went down on the seaward precipices of the reefs and planted out corals to a depth of 8½ fathoms [15.5 meters]." He added, "we really entered a new world in so far as scientific study is concerned when we got down on the submarine precipices under the breakers, and many new facts have come to light."[16]

Mayor was fascinated by the eighty-pound brass helmet, a cylindrical device with a small glass pane for viewing. In clear water, he rhapsodized, "it is a striking sight to see a beautiful mackerel with dark mottled back and golden spotted side fade into nothing but a black glistening eye when only ten feet [three meters] away." He noted, however, that a diver faced "an embarrassing top heavy tendency to upset." In addition he said that the distance of view was limited by insufficient light and that "objects loom in grey and ghostly fashion," making it difficult to judge their size. As C. Maurice Yonge has noted, use of the diving helmet was dangerous, and a diver could not stay in deep water long because of the risks. Yet, the apparatus allowed Mayor to make better observations than could be managed only from the surface. Yonge states that Mayor was the first person to use the helmet "for scientific purposes," but, in fact, William Longley and Lewis R. Cary employed it at Loggerhead Key in 1915 for scientific research. Moreover, a more primitive diving dress had been employed nearly half a century earlier in scientific collecting, and, as the ichthyologist Eugene W. Gudger said in 1918 in his report on underwater work with a diving helmet at Tortugas, the device was "not a new thing." He noted, however, that Longley and Cary deserve credit for originating "the use of the diving helmet for submarine biological work." Along with Mayor, they clearly understood the importance of underwater observations in marine ecological research.[17]

As he noted in his draft of an article on the structure of the Samoa reefs, Mayor also descended in a diving hood to a depth of 40 feet (12 meters) on the reefs off Pago Pago harbor, but in this instance, as vision was limited to 4 feet (1.2 meters) there, he was unable to ascertain much. In conjunction with his study of the water temperature and in keeping with a notion previously posed by Vaughan, Mayor believed that he had confirmed his hypothesis that diminished light affects coral growth more adversely than do decreases or increases in temperature, a finding that ran counter to James Dwight Dana's argument. Given the fragile state of his health, Mayor may have acted in a foolhardy manner by making those risky dives, but courage and a compelling desire to advance knowledge of coral reefs outweighed his sense of danger. Once again he used grids to count the number of coral heads, species of sea fans and sea whips, the sea cucumber *Stichopus chloronotus*, and the blue sea star *Linckia laevigata* in several 24-foot (7.3-meter) squares.[18] He also studied other matters that he would pursue further during a return to Samoa in 1920.

For reasons unknown, Mayor did not follow his plan to go "around the world" to study the alkalinity of seawater. In part, he may have decided against the plan because of his health, although it seemed to hold up well while he was in Samoa. Or, since he had received word of a powerful hurricane that had struck Tortugas on September 9–10, 1919, he may have considered it necessary to return to the United States sooner. After returning to Princeton, Mayor wrote a brief article on the growth rate of certain coral species and sent it to the editor of the *New York Zoological Society Bulletin,* in which it appeared in November 1919. He also set to work on an article dealing with the effects of oxygen deprivation on nerve conduction in the scyphomedusa *Cassiopea.* The article appeared in the April 1, 1920, issue of the *American Journal of Physiology.* In addition Mayor wrote preliminary drafts of his extensive studies of the Samoa coral reefs and began a biographical sketch of Samuel Scudder, a Harvard University entomologist with whom Mayor had developed an association during his years at the MCZ. His mind was much troubled, however, because of his declining health and because of the damage to the Tortugas Laboratory by the "severe hurricane" in September. The storm had wrecked the laboratory's wharf, the two main laboratory buildings, and the machine shop; toppled the windmill; and destroyed the land snails being experimented on by Paul Bartsch. The hurricane winds, reaching an estimated 120 miles per hour (193 kilometers per hour), had once again shown the risk of keeping the marine station on Loggerhead Key. Although he ordered temporary repairs to reduce additional deterioration, Mayor decided that it would be best not to open the laboratory during the 1920 season. Permanent repairs would take more time and consume a sizable portion of the laboratory budget for the year.[19]

The decision not to operate in Tortugas in 1920 also freed Mayor to return to Samoa and Fiji and continue the coral-reef studies he now considered so important. Indeed, in his report to the CIW trustees for 1920, Mayor said, "we took advantage of the opportunity to complete the only intensive study yet undertaken of a volcanic Pacific Island and its reefs." He also reiterated his view of the expanded mission of the Tortugas Laboratory. No doubt he did so because he firmly believed in that mission, but clearly he hoped to keep the CIW trustees from focusing on the station they had approved at his urging and to which, over the last sixteen years, they had allocated around $350,000. Mayor's appeal certainly made good sense in a way, and it was persuasive: "the Department of Marine Biology is much more than a mere laboratory, for it is essentially an agency for the study of biological problems of all oceans." Now clearly sharing his view was CIW president Woodward, who told Charles D. Walcott in a letter written on May 21, 1920, that the time was right for building "an international station for marine biology." He estimated that the cost of such an enterprise would be "not less than $100,000," and it would take $50,000 a year to operate the new station.[20] It is unknown whether this information was conveyed to Mayor, but his actions suggest that he knew he had the support of Woodward for expanding the CIW's marine biological laboratory.

Noting some of the important studies conducted at Tortugas, Mayor referred to the station as "a base" for inquiries that have "often led us far afield." His argument was sound in that it acknowledged the limitless nature of marine studies, but it clearly constituted a restatement of the original aim of the proposal approved by the Carnegie Institution. In this report Mayor cleverly emphasized the official title of the CIW department, that is, "the Department of Marine Biology," and referred to the Tortugas Laboratory as though it were but a branch of that department. Five years earlier he had changed the title of the laboratory publication to *Papers from the Department of Marine Biology of the Carnegie Institution of Washington,* thus de-emphasizing the Tortugas Laboratory in the original title. That the Tortugas station was a base subject to hurricane damage; that he had touted it as an ideal location for research, especially for studies of tropical biology; and that he had advocated a relocation of the base —all of these Mayor ignored in the report. His vision of studying the "biological problems of all oceans" was nonetheless noble and forward-looking.[21]

By March 11, 1920, Mayor reported from San Francisco that he had completed the biographical sketch of Scudder for the NAS, and he told Davenport that he was ready to do a memoir of another scientist. Evidence indicates that he had also begun, or at least had thoroughly considered, a popular book on his views of Oceania, its peoples, and its cultures, to be titled "Pacific Island Reveries." Less than a week later, in company with Cary, Mills, the Vassar College biologist Aaron L. Treadwell, the University of Chicago geologist Rollin T. Chamberlin, the Cambridge University zoologist Frank A. Potts, the University of California botanist William A. Setchell, and the University of California bacteriologist Charles B. Lipman, Mayor embarked for Samoa and Fiji. Nearing Pago Pago on April 2, he wrote to Harriet that he would work on the Samoa and Fiji reefs until July 26, then depart for Honolulu to attend a meeting of the Scientific Congress of the Pan Pacific Union and to study at the Bishop Museum before leaving for home on August 23.[22] Thus, he would conduct research around the islands for a period of nearly four months.

Eager to get back to Fiji after being prohibited from going there in 1919, Mayor and his colleagues left for that island on April 9. They remained there for a month, but not without disappointment. Discovering that a storm had washed away all of the sea fans planted by Cary and one-fourth of the corals he had planted in 1918, Mayor told Harriet that it had "hardly paid to come back here." He did observe a significant decrease in living reefs in the harbor of Suva. They had been thriving when he saw them in 1898, but in 1918 and 1920 he found that most of them had been killed by mud and sewage. Mayor was much more successful at Samoa, where, joined by William Longley on July 5, he again ventured into the deep in a diving hood, alternating with Longley, who was doing a comparative study of reef fishes. Once again Mills continued to drill into reefs for core samples. Potts studied the growth rates of selected invertebrates, while Treadwell focused on marine annelids. It was certain to the enthusiastic leader Mayor that this research would result in "the best volume yet published on a reef of the Pacific."[23]

During this expedition Mayor arranged a trip to Rose Atoll. In a letter to Harriet he stated, probably correctly, that the isolated atoll "has not been visited by any scientific man since the Wilkes Expedition in 1839." Although he spent only two days at that site, Mayor "got some very important scientific facts" that, in his judgment, helped to "explain the geological history of the whole group [of Samoa reefs]." In a report on the atoll, he noted that, contrary to a Wilkes Expedition map and another drawn in 1873, it really consisted of two islets: Rose, approximately 240 yards (219 meters) long and 200 yards (183 meters) wide; and Sand, much smaller. Sand Islet, he noted, contained many shell and coral fragments and no vegetation and stood only 5 feet (1.5 meters) above high tide. Rose Islet, however, stood 11 feet (3.4 meters) above high tide and served as a nesting site for noddy terns and sooty terns. Also present were boobies and numerous "small, brown and gray" rats. Mayor described the rim of Rose Atoll as smooth and hard, which led him to declare that it was not a fringing reef. Large boulders of *Lithothamnion* (=*Mesophyllum*), a calcifying alga, on the rim edge convinced him that the island had not risen but that the sea level had fallen. He believed that a higher ocean had "planed" the rim down by 8 feet (2.4 meters) over the centuries. Mayor found no trace of lava, as Charles Wilkes and Joseph Couthouy had claimed they saw in 1839.[24]

After completing thorough studies of the Samoa reefs in 1919 and 1920, Mayor was in a position to offer valuable scientific information on their structure and ecology, and he drafted three major papers on the subject, though they did not appear in print until 1924. All of them were published in *Papers from the Department of Marine Biology of the Carnegie Institution of Washington.* In his paper titled "Structure and Ecology of the Samoan Coral Reefs," Mayor did much to advance methods of looking at coral reefs, and he contributed significantly to the debate over the formation of various types of reefs. He correctly noted, for example, that "all of the islands of American Samoa are surrounded by a shore bench which is about ten feet [three meters] above present sea-level," and this phenomenon led him to suggest that it was a result of the lowering of sea level rather than of elevation of the islands. He also noted, however, that many other islands in the Pacific and in the Atlantic appeared to have "emerged above present sea-level."[25]

As a result of his Samoa reef studies, Mayor concluded that Darwin had correctly observed that fringing reefs become barrier reefs after a long period of subsidence around an island but had confused the types by failing to specify how deep and how broad the channel must be between them. Since all reefs begin as isolated patches of coral, Mayor argued that fringing reefs form as space between the patches, leaving only small tide pools but not lagoons. Thus, while Darwin was correct in the main, he erred in some specifics, said Mayor. But, he added, "Darwin labored under great disadvantages," for he saw only a limited number of atolls and confined his observations mainly to the Society Islands. Moreover, while Darwin assumed a constant rate of coral growth, stated Mayor, that assumption had since been proven wrong. Still, he credited Darwin with making keen observations about the formation of coral reefs.[26]

In a second paper on South Pacific coral reefs, Mayor again addressed the question of the cause or causes of stability of the "floors" of lagoons, challenging once more the view of John Murray and Alexander Agassiz that dissolution of limestone in lagoons accounted for their excavation. Following Wayland Vaughan, Mayor argued that the amount of limestone dissolved was relatively minimal, and he hypothesized that a ridge of *Lithothamnion* (calcareous algae) builds up at the seaward edge of a reef flat and eventually becomes high enough to keep waves from destroying the inner part of the reef, thus helping to form a lagoon. In his view, fragments of dead organisms in the lagoon, while occasionally agitated by storms, are not washed out of the lagoon as large pieces. Instead they continue to break up as a result of boring organisms, rather than dissolving in seawater. To test this hypothesis, Mayor repeated the procedures he had followed earlier at Tortugas and concluded that the amount of material carried away is balanced by additions, including new corals.[27] To Mayor, the Murray-Agassiz explanation was simply wrong.

Mayor's third paper dealt with the rate of growth in several species of Pacific corals. Since 1890 studies of Pacific corals by British scientists and of Atlantic corals by Vaughan had added to knowledge of coral growth, and Mayor advanced the subject considerably by conducting seven sets of experiments between 1917 and 1920. In each case he removed a segment, or head, of coral and then photographed, weighed, measured, and tagged it. Then he fastened it to an iron stake or set its base in concrete and returned it to the water. In the first two experiments he took the specimens from shallow water and returned them to the same place from which he had taken them. In the third experiment he obtained the specimens from a depth of 2 to 3 fathoms (3.7–5.5 meters), by use of a diving hood, and returned them to the same place. The fourth set of coral specimens came from the "breaker-washed edge of a reef patch." Again, after obtaining data he placed them back on the edge. Sets five and six were specimens of the massive corals in the genus *Porites,* and the procedure was the same.[28]

Each year thereafter he removed the specimens, which were labeled with brass tags; obtained data to ascertain rate of growth; and took photographs for additional study. The seventh set of experiments was especially ingenious. Obtaining twenty-four specimens of several species, either in "highly agitated shallow water . . . [or] in deep, quiet water," Mayor sawed each in half, embedded the base in quick-setting cement, and placed one-half of the total number in water that fell to eight inches (twenty centimeters) at low tide. He placed the others in "quiet water" at various depths, ranging from "moderate" to 5–8.5 fathoms (9.1–15.5 meters). When later removed, the corals showed variations not only in rate of growth but also in form, the latter factor reflecting the influence of depth, light, temperature, and wave action in shaping various forms of the same species of coral.[29] Mayor planned additional experiments in order to determine the influence of location on the morphological plasticity of scleractinian corals.

Perhaps the most venturesome, and certainly among the most informative, aspects of his study of coral growth came from his underwater investigations. "Using a diving-hood," Mayor wrote in the third paper, "I have gone down on the seaward edge of the Aua Reef when the breakers would permit such action, and made a study of the abundance of the various genera of corals." He added that he had planned to use the same grid approach he had used in shallower waters, "but the extreme irregularity of the surface [of reefs] made this impossible." That difficulty, he noted, was exacerbated "by the surges which tended to throw one off one's feet, or into intricate caverns from which it might be difficult to escape." It is almost certain that Mayor never told Harriet of the risks he took underwater, though he may well have informed her of the joy he received in walking "around among the coral heads at depths down to six fathoms [eleven meters]" while writing "elaborate notes" on "a white board." Although unable to be as precise as he wished, Mayor could, and did, make counts in roughly defined areas. Of the scleractinian genera he counted, he estimated that 87 percent were *Acropora* and ten percent were *Pocillopora.* The remainder consisted of the hard coral *Porites,* the anthoathecate hydroid *Millepora,* and a genus of alcyonarians, or soft corals. His observations led him to conclude that *Acropora* species grow the fastest, especially toward the seaward edge.[30]

Meanwhile, the Harvard physical geographer William Morris Davis, with whom Mayor had disagreed earlier, read Mayor's most recent paper on the

formation of Tutuila coral reefs, and on October 5, 1920, he penned a long letter of objection to Mayor's hypothesis. Contending that a fringing reef could
not develop successfully in a lagoon, he stated his belief that the reef identified
by Mayor as a fringing reef at Tutuila must be an "inner flat." Davis also objected to Mayor's "criticism of Darwin, where you say that the submerged barrier reef of Tutuila was not developed from a fringing reef." In fact, said Davis,
"you misunderstand Darwin." Mayor quickly wrote a reply to Davis and sent a
copy of it and the Davis letter to the geologist Rollin T. Chamberlin for his
opinion. In his letter to Chamberlin, Mayor referred to Davis's "peculiar psy-
chology in insisting upon calling Barrier reefs fringing reefs," and he said, "the
truth is he [Davis] *knows* he is wrong and is simply stirring up mud to cloud the
issue." Irked by Davis's criticism, Mayor offered a sharp reply. He asked Davis
whether he could name any richer fringing reef other than the one that was
already fused in places with the barrier reef west of Suva harbor. The outer edge
of the fringing reef, he noted, consists of dense clusters of fast-growing species
of coral, but an especially strong current that flows between the fringing reef and
the barrier reef prevents complete fusion. The barrier reef, he added, had arisen
in situ. Drawing a clear diagram of the phenomenon, he sought to show Davis
its nature.[31]

Mayor went further in his effort to show that Davis was, in his opinion,
offering a purely speculative view of the matter. He stated that if Davis were to
study the reefs underwater in a diving bell, he would find evidence inconsistent
with his speculation. In such a case, as Mayor knew from experience, Davis
would be convinced that the current was so strong that he could hardly stand,
and he would see densely crowded growths of *Acropora* and other coral species.
Moreover, he asked how Davis could refer to an "alluvial flat" when "its *outer
edge* [is] *so steep.*" He could not agree with Darwin or anyone else, he said, who
calls a barrier reef a fringing reef. Aristotle, said Mayor, believed that flies have
four legs, but a modern biologist can not say that four legs equal six legs. In the
same way, added Mayor, a fringing reef is not a barrier reef. The analogy was not
especially pertinent, but it was, as Mayor hoped, sufficient to irk Davis. "Pardon
this abruptness," wrote Mayor, "[but] I know you like a hot fight . . . ; so come
back at me."[32]

In a reply to Mayor's letter of October 9, Chamberlin stated that he found it
"a bit strange" that Davis "should attempt to instruct you upon the conditions
of growth of fringing reefs" since he had no experience "with the diving helmet."
Davis was engaging in "arm chair geology," wrote Chamberlin, adding that it
appeared to him that Davis had begun with "partial information" and was
ignoring "a lot of actual factors which play an important role in the real case."
Chamberlin commended Mayor for refusing to "confuse fringing reefs and

barrier reefs." Coming from a distinguished geologist, Chamberlin's words of encouragement meant much to Mayor.[33]

Study of the formation of coral reefs had clearly intensified by this time, and Mayor found himself deeply engaged in the debate over their origin. Interest in the topic was not new, of course, since Charles Darwin had published *The Structure and Distribution of Coral Reefs* in 1842. Darwin's insightful work later gained important support from James Dwight Dana, a leading American authority on the subject. Continuing his interest in the subject, Darwin issued a revision of his book in 1874. Especially intriguing to Darwin was the origin of atolls, which, contrary to popular opinion, had not developed along the rims of submerged craters of volcanoes, he stated. Contending that every reef had originally formed on a foundation of rock, Darwin hypothesized that reefs continued to grow upward until they reached the surface. In the case of an atoll, reasoned Darwin, a reef formed on a submerged basaltic base. As the volcano began to subside slowly, the reefs continued to grow on the seaward edge, building on themselves always to the ocean's surface. As the center of the volcano generally subsided more quickly than its sides, the slow process of reef building could not keep pace there, but a ring of reefs on the outer perimeter continued to thrive and eventually surrounded a lagoon.[34]

The same process, according to Darwin, manifested itself in the development of reefs along the shores of larger land masses. Originally those reefs fringed the islands and continents where conditions were conducive to their growth. As shores subsided faster than the growth of reefs on their seaward edge, the reefs became separated from the land mass by a lagoon. Thus, maintained Darwin, these became barrier reefs. Darwin believed that deep drilling would establish that every reef was formed on a rock base, and he expressed a wish for such a test of his hypothesis. Between 1896 and 1899 the Royal Society of London funded three drilling projects to verify Darwin's view. Even at a depth of 1,114 feet (340 meters) on the third probe, however, scientists found ancient coral material. Of course, Mayor also attempted to test Darwin's supposition, and his second drilling probe resulted in a find of basaltic rock at the base of the more recent reef through which the probe was made. Conclusive evidence remained for a later time and better equipment, however, which did not come about until 1952, when scientists were able to bore into the Eniwetok atoll in the Marshall Islands. On two occasions the probes reached depths of four-fifths of a mile (1.3 kilometers), and each time they found a basaltic base, as Darwin had guessed.[35]

In the meantime, from 1872 to 1876 Britain's *Challenger* Expedition had conducted significant oceanic studies. One of the members of the *Challenger* team was John Murray, who devoted considerable attention to coral reefs. Later knighted, Murray developed ideas about reef formation that, along with augmentation by Alexander Agassiz, held sway until T. Wayland Vaughan, Alfred G. Mayor, Reginald A. Daly, and others began to raise objections to their views. As a result of his studies around Tortugas and the South Pacific islands, especially around American Samoa, Mayor had become convinced, of course, that Murray and Agassiz erred in assuming that "sub-marine solutions," or chemical reactions, resulted in the deepening of lagoons. Along with Vaughan and Daly, he believed that Murray and Agassiz were also wrong in positing that "calcareous deposits" formed into hard substrates on ocean floors or as platforms or shelves along the sides of submerged volcanoes and thus became the bases of coral reefs.[36] One of Mayor's contributions to knowledge of lagoons was to demonstrate the inadequacy of the Murray-Agassiz hypothesis.

Mayor failed, however, to refute the Darwin-Dana view on the role of subsidence in the formation of atolls and barrier reefs. As noted by a later authority, Norman D. Newell, modern explanations of how coral reefs form consist of "a synthesis of many ideas once regarded as irreconcilable." Newell states "that no *single* explanation can account for *all* coral reefs." As Newell observes, even Darwin's hypothesis, though now considered basically correct, is "incomplete . . . because it did not take into account the comparatively recent and great fluctuations in sea level produced by the waxing and waning of the Pleistocene continental glaciers." In fact, Mayor followed Daly's glacial-control theory too closely, and both he and Daly rejected any major role for subsidence.[37] Mayor could have profited from entertaining some of Davis's useful explanations about the nature of reef building. Still, Mayor's views stimulated thought on this important subject, and his investigations served to fill gaps in the field of reef research. Moreover, as C. Maurice Yonge has observed, along with Vaughan, Mayor helped to extend "coral morphology and taxonomy to embrace more dynamic aspects covering function and ecology." Mayor explored the impact of temperature, sedimentation, depth, wave action, and freshwater on coral reefs, and he generated useful data on the growth rate of corals.[38] Given the few years he actually devoted to the study of coral reefs and corals and his avid interest in continuing his inquiries, he would likely have made other significant contributions to the field, but his health was to fail him.

Mayor was now experiencing pain in his throat and needed a boost to his spirits. Despite the pain, he wrote up his report for the 1920 season, which included nothing from the Tortugas station since the laboratory was not officially opened for research during that year. The report focused on the work done by him and his team in Samoa and Fiji. There could no longer be any doubt that Mayor was infected with tuberculosis, though oddly, he apparently believed that the infection had occurred during the "last field season." In a letter to Mayor on October 30, 1920, Woodward offered his sympathy and assured his friend that new ways had been developed for "combating this insidious disorder." The condition of Mayor's throat was so bad by mid-November that the ailing biologist could speak only in a whisper. Nevertheless he remained optimistic throughout the ordeal, assuring one of his correspondents on November 19 that he was "getting better steadily." Two weeks later he again claimed that he was "all right," but he complained that his throat was "healing slowly." He added, in jest, "I am thankful . . . I wasn't born a giraffe." Soon thereafter he told his old friend Davenport that he was "getting on in good shape and expect to go to Tortugas as usual." Again he attempted to make light of a serious situation, telling Davenport that his wife was severely restricting his activities. "Harriet has put me in Jail, cell no. 13," he jested. Inconsistent with his belief that he had not been infected before the previous summer, he now said that in all probability he had caught the disease from Harriet. "It certainly is a joke on me," he added, "to come back from Samoa with T.B. in my throat instead of getting leprosy, elephantiasis, or a leg bitten off by a cannibal, shark, or crocodile."[39]

Mayor continued to work on his Samoa reef papers during January 1921. He sent specimens of fossils and corals and bottom samples to Vaughan, who analyzed and wrote a report on them. Vaughan had not worked directly with Mayor in five years, but Mayor had not forgotten the ability of his old acquaintance. By late May, Mayor's condition had improved, though not sufficiently to warrant a return to Loggerhead Key. He went anyway. On June 27 he wrote to Harriet that his throat was somewhat better, but he admitted, "I literally lie down all day unless a raft of mail hits me and keeps me writing all day." Since Brantz was with his father at Tortugas, Harriet cautioned Mayor about guarding their son from the disease, "whether you [have] ceased coughing or not." She was obviously aware by now of the infectious nature of the dreaded disease. Mayor left Tortugas on July 25 and headed to Maplewood to visit his stepmother for several days before returning to Princeton.[40]

There was relatively little for Mayor to report to the CIW regarding his own studies at the Tortugas Laboratory, but he did make some observations about the fate of the loggerhead turtle, *Caretta caretta,* on Loggerhead Key. When he first visited the key in 1897, he noted, "on practically every night during the breeding season from May until August at least one female Loggerhead turtle crawled up upon the sands to lay its eggs, but in 1921 not a single turtle" appeared. Mayor said that fishermen had told him in 1897 that as many as forty females had deposited eggs during a single night around 1860. He lamented "the extermination" of the loggerheads on the key and expressed fear that the general "destruction of the littoral element of marine life" would increase, mainly because of "the oil-burning steamer." He added, "everything between tides is thickly covered with a mass of black crude oil." Although he was genuinely concerned over the pollution of the once-pristine waters, Mayor also intended to convey a message: the laboratory on Tortugas must be viewed as a "temporary" station and the CIW's laboratory should be viewed not as a "fixed station" but "as an agency for the study of problems of the oceans."[41]

Mayor also used his spare time to conduct another small study during the brief season on Loggerhead Key in 1921. It appeared in the *Papers from the Department of Marine Biology of the Carnegie Institution of Washington* in 1922 under the title "The Tracking Instinct in a Tortugas Ant." The ant he studied, *Monomorium destructor,* had somehow been introduced on the islet, and it was "a great pest in the wooden buildings of the Tortugas laboratory." Indeed, the infestation of these "voracious" ants had compelled Mayor to suspend beds from the rafters and to apply corrosive sublimate, or mercuric chloride, to the suspension ropes and to the legs of furniture. As his son Brantz later recalled, one of the scientists, a newcomer who allowed his sheet to touch the floor, was stung so badly by a swarm of them that he lapsed into unconsciousness for a while. Mayor was intrigued by this "tracker-ant" and conducted a series of experiments to study its pursuit of food and its method of alerting members of a colony to the presence of food.[42] He was unaware of pheromones, or chemical secretions in ant species, of course, but he nevertheless made some useful observations. Along with the comments on oil pollution, this study incidentally pointed up the fact that problems on Loggerhead Key were not confined to the vagaries of weather.

Unable to remain idle for long, Mayor turned his attention to his book manuscript "Pacific Island Reveries," bringing it to completion either in 1920 or, more likely, 1921.[43] He probably intended to polish it further but did not get the chance to do so. Along with various comments in his personal papers and the articles on Oceania he had published in 1915 and 1916, the manuscript rounded out his cultural and racial views. As was the case with a considerable number of his contemporaries, Mayor's views constituted a mixture of prevailing stereotypes, social Darwinism, and eugenics theory. They also represented the thoughts of an able scientist who, like many of his peers, believed that a good biologist was ipso facto a good student of culture and race. Indeed, Mayor committed the error of viewing scientific inquiry in zoology as freely applicable to human societies. It is now readily apparent that Mayor formed generalizations about human races and cultures based on flawed analogies between them and the natural world. It is also clear that his views suffered from inconsistencies that would have been unacceptable to him in his scientific work. Equally obvious are the dominating influences of the era in which he lived, but also notable are Mayor's reservations about forcing Western culture on others.

In Tahiti, Mayor had written in 1915 in his series on that French-controlled island, "the beautiful is wedded to the grand." There, he added, one can see "the sylvan setting of the waterfall where rainbows float on mists among the tree ferns." Enthralled by the beauty of the land, he said, "one is never away from the murmur of rippling water, as the mountain streams splash among moss-covered boulders." The natives were equally beautiful, but the French had managed "to curse" their culture by inflicting "civilization" on them. Though Mayor referred to the barbarous customs once followed by the Tahitians, he believed that one should not be "over harsh in condemning" them. Certainly, in his judgment, the introduction of Christianity into Tahiti had checked the practices of infanticide, human sacrifice, and intertribal war, but Mayor also believed that "the adoption of Christianity [had] contributed to the increase of certain fatal diseases, notably tuberculosis," by compelling the natives to wear "dirty European clothing" and by forcing them "to huddle together in unsanitary, ill-ventilated 'shanties.'" Equally bad, said Mayor, was the imposition of Western standards on the Tahitians, for this action had led to "the destruction of almost all that once was theirs" and ruined their self-respect.[44]

Mayor's series of articles on Fiji resonated with similar views. Of the beautiful "mountain valleys" of that land, he said, "time can not efface [them] from the memory." In those valleys, he wrote, flourish "the rich brown stems of tree ferns crowned by emerald sprays of nature's lacework," and in the lush forests, "through a break in the canopy a furtive beam of sunlight penetrates to gild the greenness of the shade." To him, however, as a human being the Fijian was not as beautiful as the pure Polynesian, for his "eye lacks the languid softness of the Polynesian's and is small, swine-like and often bloodshot, imparting a cruel aspect to his visage." Gradations in physical comeliness are thus obvious in Mayor's notions of racial types. Following the prevailing view of his era, Mayor erroneously believed that the dark skin and kinky hair typical of these Melanesian people signified African origins. Therefore, he placed them lower on his scale of human beauty. Yet, to Mayor the Fijians were to be admired. To him, "the native grace and unconscious dignity of these superb people" could not be denied. Nevertheless, in his view they lacked the stamina of the pure Polynesians and were more subject to diseases.[45]

Noting that the customs of the Fijians—indeed, all the natives of Oceania—were incorrectly labeled by Christian missionaries as "heathen" and "pernicious," Mayor said, "it is a common belief that the savage is more cruel than we [Americans]," but he viewed those "cruelties" as no worse "than the lynching or burning of Negroes," which, he observed, is "common in America." In our own country, he added, "despite our mighty institutions of freedom . . . there remain savages among us." Mayor rejected the belief that his country, or any other great power, could civilize any "primitive race," for "under our domination the savage dies, or becomes a parasite or peon." However, in his view the imperial powers had done a great injustice to the Fijians and other peoples of Oceania by introducing tuberculosis and by devaluing their art, which, he averred, represents "the highest expression of their intellectual life." The art of the Fijians, Mayor aptly observed, serves as "a record of their history and their conception of the universe." Mayor called for an end to exploitation of the Fijians and expressed a hope that, by leaving them to themselves, they might eventually "attain civilization." Torn by conflicting desires, Mayor wanted to leave them to themselves while "at the same time making them more "Western." If they did not follow the Westernizing course, they were doomed to extinction, but if they surrendered completely, they would no longer possess a culture of their own, thought Mayor.[46] Consistency could not be a jewel to Mayor in deciding the fate of the South Sea Islanders.

Yet, there can be no doubt that he sincerely believed that domination of the peoples of Oceania had been bad for them on the whole. He held the same view toward natives of New Guinea and Aboriginal inhabitants of Australia. After visiting Australia, southeastern New Guinea, and Java, Mayor had, of course, drafted three articles on the natives of those lands: "Papua, Where the Stone-Age Lingers," published in 1915; and "The Men of the Mid-Pacific" and "Java, the Exploited Island," published in 1916. Regarding the island of New Guinea, which he visited briefly in 1913, Mayor contended that Europeans had paid little attention to the land because there "the heavy air . . . flows lifeless and fetid over the lowlands as if from a steaming furnace." Even though he considered the indigenous people of New Guinea to be inferior to Polynesians, he believed that Europeans should educate them "in the production and sale of manufactured articles" and encourage them to resume their traditional arts and crafts in order to enhance "that self respect and confidence . . . which the too sudden modification of their social and religious systems is certain to destroy." The missionary schools, he asserted, were devoted to converting natives to Christianity and to teaching them the traditional European school subjects. Mayor called for the establishment of secular schools designed to teach manual training to adolescent Papuans. In his opinion, the secular schools would not "interfere with the religious teaching received from missionaries." It did not matter, he added, whether Christianity was "true or false" since the natives of New Guinea "are destined to be dominated by Christian peoples."[47]

Mayor's knowledge of the land was largely restricted to the southeastern segment of the island, and Mayor praised the British administration of that area from 1884 to 1906. However, he was less pleased with the policies of the

Commonwealth of Australia, which assumed control of the region in 1906. Although he obviously admired Papuans and made some excellent photographs of them, he was less certain about the value of independence for them. Describing the Papuans as "dark chocolate" in color, "weakly made" in body, and mentally limited, Mayor nevertheless believed that they faced "the sunlight of happiness" in the years ahead under the leadership of their imperial masters.[48]

Mayor saw the same attributes of race in the Aboriginal Australian, whom he referred to as "among the lowest of existing men." He viewed him as in "striking contrast to the finer races of the Pacific." Indeed, Mayor contended that the Aborigine could not rise "to the intellectual level of the natives of the Pacific Islands"—presumably, in his opinion, not even to that of the Papuan. As with other dark-skinned peoples, the Aborigine, in Mayor's opinion, was characterized by "little eyes [that] glitter suspiciously from deeply sunken orbits hidden under unkempt locks of matted hair." Mayor's typification of the Aborigine digressed even further, holding that one forms "a demon-like picture as he skulks silent and snake like through the thickets."[49] Mayor was unable to see the inconsistencies in his arguments and to employ the same standards toward the study of humans that he applied to the study of cnidarians. He could accept variations of color and form among the latter as indications of beauty, but among humans color variations to him were indications of gradations in superiority.

Splashes of nature's beautiful hues and tints were also important to Mayor, and he delighted in painting word portraits of the physical world, a sense he had acquired during the days of his boyhood and perfected in his exemplary illustrations. The vast oceans and their life and the lovely landscapes he observed evoked in him a deep appreciation of the natural realm. Thus, in his "The Islands of the Mid-Pacific," he spoke of the beauty of the "olive and yellow-greens, mauve and purple-browns [of] the living corals," and he lauded the great diversity of starfishes, sea anemones, and giant clams in the waters of those islands. Java was "the garden *par excellence* of the tropic world" and an edifice of beauty, "glistening and brilliant . . . with the sunlit sea around it, a shimmer of turquoise and emerald set in the everlasting blue of the [ocean]."[50] It is no wonder that Mayor loved to visit these islands; they were places of unparalleled beauty.

Of course, Mayor hoped to make money from the publication of his "Pacific Island Reveries," but he also wrote the work as lament to the waning of a fascinating culture. Polynesia, he said, "still lingered in the autumn of its decline" when he made his first visit to Oceania, but to him it had "now vanished forever into the night of things that were." In the Marquesas Islands, for example, "the moral and physical corrosion of [a] once proud race" was evident to him.

Yet, in contradiction to his main argument about the unsavory influence of Western intruders, Mayor contended that the decline of Polynesia was also a consequence of "the age-old communism" embraced by its peoples. In his judgment, the "archaic socialism" of Polynesian culture "stultified individual ambition" and lessened the "mental power and moral stamina" of its people. Obviously embracing the ideal of the capitalist system and almost certainly concerned over the implications of the recent Bolshevik triumph in Russia when he penned those views in 1920 or 1921, Mayor had softened his criticism of the Western powers. He also expressed greater doubt about the capacity of the peoples of Oceania and Australasia to govern themselves effectively, for he believed that they were all "childlike." Even the Polynesians, in his view, could reach no further than a mental age of fourteen. Such was the outcome of the process of evolution by natural selection, he added. Still, something could be done to save the remnant, and that was a service that "another type of missionary" could perform. The new type of missionary would not be a "religious zealot" but rather a teacher of manual training. The schools run by the Christian missionaries, he said, are "stupid" and teach nothing useful to "these docile . . . and amiable children."[51]

Mayor was also concerned about the potential of Japan to dominate Oceania. In fact, he expressed a fear that Japan might endeavor to dominate not only the whole Pacific but also the world. "Truly in the Pacific," he wrote, "we behold the sea of destiny, the ocean upon which must be fought out the vast question: who shall eventually dominate the world, Asia or Europe?" Mayor had visited Japan during the voyage of 1899–1900, and he had come to conclude that, although the nation had adopted "the modern sciences of Europe," it had clung to its "national obsessions" and its notion of being "the chosen people." Warning that Japan might "plunge itself into fatal conflict with the Gentile races," Mayor spoke of a "racial trait" and a "spider-like quickness" of the Japanese that must be watched carefully.[52] Mayor's views once again reflected elements of perception while indicating racial stereotypes. They also reveal the way in which an excellent scientist allowed himself to mingle his own cultural bias with the theories of evolution and eugenics to grade races. His use of science was in fact a misuse, but it was common in his day.

By the time Mayor had completed "Pacific Island Reveries," his strength was greatly diminished, and his throat once more became so sore that he could hardly speak. In early October 1921 a physician advised him to spend the winter in a sanatorium in Tucson, Arizona, though he told him that there was "very little T.B." in his lungs and a "surprising improvement" in his throat. In fact, three weeks later Mayor reported to Charles Davenport that the physician had predicted "a complete recovery within the year." Yet, Mayor harbored misgivings, and in a letter to Davenport he said, "a *nice* degenerate family we are, with both Harriet and myself T.B.ites." He expressed concern that "all the four kids having the gene on both sides will get it [tuberculosis] and drop like flies!" This was a matter, he added, that "shows how difficult it is to *apply* Eugenics." He noted that none of his or Harriet's forbears had suffered from tuberculosis. Some of them had succumbed to alcoholism, he said, but "none keeled over with T.B." Continuing to believe in the genetic cause of tuberculosis, he concluded, "The only time to *really* judge whether a person is fit to marry is after they [*sic*] are dead."[53]

Although he was rightly worried over his health, Mayor continued to plan for another season at Tortugas in 1922 and yet another expedition to the South Pacific. On December 22 he wrote to Joseph A. Cushman, an authority on foraminifera, approving an additional allotment of $520 for the cost of illustrations in a paper Cushman was preparing for a Tortugas Laboratory publication. He added, "I will be at Tortugas from about May 15 to July 25, 1922, and wonder if you care to come." Sometime soon thereafter Mayor entered a sanatorium in Tucson, Arizona. He continued, however, to believe that he would be able to open the laboratory at Loggerhead Key in May and go to Tahiti during the fall. In a letter to Cushman on March 5, 1922, he said that he planned to make borings in the coral reefs at Tahiti and "tackle *barrier reefs,* having done the fringing reef problem in Torres Strait and Samoa." Not long thereafter, however, his condition worsened, and on the last day of March he wrote to his close friend Stewart Paton that he had contracted a case of influenza that "gave me a great smash," resulting in "severe bronchitis . . . and deeply blood-stained sputum." He admitted that the attack had left him weak. "Just how many weeks, or months, this attack will set me back," he lamented, "I do not know." Ever optimistic, however, Mayor stated that he "expect[ed] soon to be over it in good shape."[54]

By mid-April, Mayor was feeling better. In fact, he was well enough to grouse over the recent decision of the ever-petulant Wayland Vaughan to complain to J. Stanley Gardiner that Mayor should not have invited Charles Lipman to study "the bacteriology of the sea water of Samoa." Mayor correctly noted that Lipman, a University of California scientist, was eminently qualified for the task and that he (Mayor) had invited him to continue his work at Tortugas. Vaughan contended, however, that "Lipman's work is not worth consideration," and he insisted that Lipman "knows nothing of the literature of the subject." Mayor wrote to Vaughan to tell him that he was being unfair, but the obstinate man would not relent. "I fear that jealousy enters into the matter," Mayor told Walter M. Gilbert, the CIW secretary. Noting that Vaughan "claims to be the only authority upon the percipitation [*sic*] of calcium carbonate in sea water," he added that Vaughan "seems to think that the subject began and ended with him." Mayor concluded, "we all know Vaughan's idiosyncrasies, but in this case . . . he has gone beyond the limit of justice, or of common sense."[55] Despite his serious condition, Alfred Mayor had lost none of his zest for fairness.

He was discouraged, however, that the stay in the sanatorium had done little to improve his condition. In mid-March he ran a constant fever and suffered fits of violent coughing, but on April 17 he wrote to his personal physician in Maplewood that although he wanted to be examined by him, it was "really impossible" as he planned to leave Tucson and go directly to Tortugas in early May. The violence of his coughing spells had decreased and the tubercles in his throat had disappeared, he noted, but his vocal cords were not much improved. Also, his body temperature continued to run above normal. An examination by his physician would have to wait until the fall, after he returned from Tortugas, he said. Sometime early in May, Mayor left the sanatorium and headed for Tortugas. The trip was taxing, and according to a letter from his stepmother, Mayor received "another set-back." Though gravely ill, Mayor tried to carry on and was at Tortugas by May 10.[56]

It was clear to Harriet that Alfred was unlikely to survive or, if he did, to be able to continue his work, and in a letter to her husband on June 11, she said that she was looking for ways to "greatly increase my earning capacity." The ever-devoted wife added, "I am longing to make things more comfortable for you." Alfred endeavored to remain optimistic, and on the same day he penned a letter to Harriet, saying that he was taking chaulmoogra oil for his throat and getting "plenty of fresh vegetables and good meat from Key West." He admitted, however, that he had been "struggling with high fever for a long period" and that he was continuing to lose weight. "On the whole I don't like the outlook but am all in the fight," he added. Five days later, however, he wrote to his "Dearest Old Chum" that he would like for her to "come down and spend about a week at Tortugas and see your old man home." Trying to allay her fears, Mayor said, "the fever is reduced, and I feel better." Harriet soon departed Princeton for Key West, where she was scheduled to meet the *Anton Dohrn* on June 26 and be transported immediately to Loggerhead Key.[57]

Meanwhile, on June 6, the first group of researchers arrived on Loggerhead Key. Among the four in that group was William H. Longley, who later reported that, despite his illness, Mayor met them at the dock, as he normally did. Longley found the ailing director "pale and thin," and he noted that earlier during the day on which the party had arrived, Mayor had choked badly while taking chaulmoogra oil. Over the next few days Longley observed that it took Mayor an hour to eat, often coughing, choking, and vomiting during the process. As the summer temperatures soared, Mayor's strength waned even more. Bouts of diarrhea plagued the deathly ill man, and breathing became more difficult for him. Mayor continued to write letters, but he had insufficient energy to do much else. On June 24, after observing that Mayor had suffered through a fitful night of labored breathing, Longley took him a glass of water and expressed sympathy over his suffering. Mayor replied, "I'm not complaining. . . . I'm fighting." Mayor told him that he was cheered by the prospect of seeing his beloved Harriet two days from that time.[58]

At approximately 2:00 A.M. on June 25 Mayor arose in a delirium, and Longley calmed him until he was able to return to his bed. After daybreak Mayor arose, and the assistant treated his throat with chaulmoogra oil, which caused Mayor to faint. Longley insisted on staying with him until Harriet arrived, but Mayor objected, saying that he would not allow any interference with Longley's scientific work. Longley decided to humor Mayor by going out onto the water at 2:30 P.M., but he stayed close to shore so that he could be signaled easily. Longley returned two hours later and asked the assistant to check on Mayor. When he could not be found, Longley headed for the water's edge, fearing that Mayor had gone there to wash his underclothes and had fainted. Indeed, he had, and Longley found him "lying face down in the water."[59]

The end had come in the sea, but the report of the coroner's jury, issued soon after Longley and Mills transported Mayor's body to Key West, indicated only a small amount of water in Mayor's lungs. The report was based solely on the opinion of the undertaker. In any case, as Longley later reported, the coroner's jury ruled that "death was due to heart-failure and general debility contingent upon his tubercular condition." Longley privately suspected that Mayor had drowned, but he could not be certain. Less than three hours after transporting Mayor's body to Key West early on June 26, Longley met Harriet as she stepped onto the Key West dock. He gave her the sad news and told her of the coroner's conclusion. Harriet believed that Alfred had died of natural causes, and Longley later urged others to accept that view. In his account Longley said, "Dr. Mayor was no quitter."[60] Indeed, he was not, and even during his last days he had struggled admirably to perform his duties as director of the Tortugas Laboratory.

*Epilogue*

The funeral service for Alfred Goldsborough Mayor was held at his stepmother's home in Maplewood, New Jersey, on Thursday, June 29, 1922. Obituaries appeared in several newspapers and scientific journals soon after his death, and his old friend Charles Benedict Davenport later wrote a biographical account of him for the National Academy of Sciences. While Davenport's tribute provides a good glimpse into the life and work of his old friend, it is marred by a marked emphasis on eugenics. Tributes by others, though much briefer, offer a better picture of the man. Among them is one by Charles H. Townsend. Acquainted with Mayor since he had met him during the expedition to the South Pacific in 1899, Townsend wrote in the *New York Zoological Society Bulletin* that Mayor "had an unusually lucid mind" and that his knowledge of "scientific theories was so clear that it became a pleasure to listen to his expositions." He praised Mayor for his modest demeanor and "attractive personality." Similar praise came from Frank A. Potts, of Cambridge University, in a eulogy written for *Nature:* "The energy and vitality of his body and mind, his dramatic sense, the tenacity of his memories of men and countries, the range and grasp of his knowledge, all never failed to raise the admiration of his friends." Potts also noted that Mayor had hoped to move the Tortugas Laboratory to Jamaica "and make it a truly international meeting place for biologists."[1]

By early October 1922 Joseph Cushman had organized an effort to erect a monument on Loggerhead Key in memory of Mayor. Harriet Mayor considered the monument to be a fitting tribute to her late husband and agreed to design a bronze plaque for it. Most of the scientists who had worked at the Tortugas Laboratory donated money to the cause, and most of them offered words of praise for Mayor when they sent their contributions to Cushman. For example, the curator of ichthyology for the American Museum of Natural History, Eugene W. Gudger, who called Mayor "my close friend," lauded him as a great scientist, while Charles Lipman spoke of Mayor's generosity toward him and claimed to possess "a reverential regard for his memory." The bronze tablet was erected during the summer of 1923 by John Mills, the faithful engineer and general handyman who had worked with Mayor every year since the Tortugas Laboratory had opened. In the meantime, on June 22, 1922 the Executive Committee of the CIW had passed a resolution of tribute to Mayor, noting his scientific accomplishments and praising him for "his peculiar genius for the organization of research, his extraordinary enthusiasm, his insight into the needs of science, and his loyal and unselfish devotion to its cause." In October of the same year the committee voted to approve payment of Mayor's salary to Harriet for the period from October through December 1922 and a regular amount thereafter to supplement her annuity income "to assure her the sum of $100 a month" as long as she remained "the widow of Dr. Mayor."[2]

In March 2004 one of the authors traveled on two occasions to Loggerhead Key to visit the site of the Tortugas Marine Biological Laboratory, and the other toured the area in October 2005. A century after the station was founded, they viewed the ruins of the facility, most of which are unlikely to have been part of the original laboratory complex erected in 1904. The most prominent of the ruins were the once-rugged concrete walls of what constituted part of a water cistern system on the northeast side of the property. Several concrete slab foundations were present as well, some of them intact, others broken up. Other remains of the station, observed in intertidal or shallow subtidal zones on the northwest perimeter of the property, included rails from the shipways, a dock piling, pieces of water pipes, and a strand of corroded steel guy wire with a turnbuckle. No buildings remain from any period that the station was in operation, the last of them having been destroyed by a fire of unknown origin in 1964.[3]

Only slightly less conspicuous at the site than the concrete cistern walls was the monument erected in honor of Mayor. Secured vertically to the northnorthwest face of a large concrete block that now shows cracks and other signs of deterioration from the hot, humid, salty, windy, and at times stormy environment, the rectangular plaque, approximately 5 feet wide by 2 feet ($1.5 \times 0.6$ meters) high, bears the following inscription:

ALFRED GOLDSBORO MAYOR
WHO STUDIED THE BIOLOGY OF MANY SEAS AND
HERE FOUNDED A LABORATORY FOR RESEARCH
FOR THE CARNEGIE INSTITUTION DIRECTING IT
FOR XVII YEARS WITH CONSPICUOUS SUCCESS
BRILLIANT VERSATILE COURAGEOUS UTTERLY
FORGETFUL OF SELF HE WAS THE BELOVED LEADER

OF ALL THOSE WHO WORKED WITH HIM AND WHO ERECT
THIS TO HIS MEMORY BORN MDCCCLXVIII
DIED MCMXXII

Etched on each side of the inscription is a torch, representing devotion to inquiry and to leading others to advance scientific knowledge.

Gone in addition to the laboratory buildings were the windmill, the aquarium, and the dock. Missing too were the coconut palms, azaleas, date palms, rubber trees, bananas, and ornamental cacti planted early during Mayor's directorship. As described in his annual report for 1905, these had been placed on the property to reduce glare, to beautify the grounds, and to ameliorate the impact of hurricanes on the buildings. At the time of the previously mentioned visits to Loggerhead Key in 2004 and 2005, the site was partially overgrown by low-lying vegetation, including prickly pear cactus, sea purslane, morning glories, and certain other salt-tolerant species. A cluster of sea grape trees grew along the northeast side of the property. Casuarina trees, planted over the area sometime during the twentieth century, had been cut down and left within previous years. Weathered logs, limbs, and stumps from the eradication of those non-native trees littered the site, magnifying the sense of destruction of what at one time had been an active and tidily landscaped facility.

Considering the vast amount of time and effort that Mayor expended in establishing, operating, and maintaining the Tortugas Laboratory, and the high hopes and dreams he initially had for it as the only tropical marine biological station in the United States, it was poignant to see it lying so completely desolate and abandoned just one hundred years after its founding. As for the grounds formerly occupied by the once-prestigious Carnegie marine laboratory, they will gradually return to a less chaotic and more natural state. Both Loggerhead Key and waters surrounding it are now protected as part of the Dry Tortugas National Park, and visitation to the island is strictly controlled. Moreover the marine environment of the Tortugas region is still noteworthy for its clear waters, rich biological diversity, relatively pristine state, and remarkable natural beauty that first attracted Mayor to it.

The fate of the Tortugas Laboratory had become a question immediately after the death of Mayor. In fact, as early as September 20, 1922, William Longley expressed concern over its future, reporting that Mayor had told him that it might be abolished. Woodward had retired as president of the Carnegie Institution in 1920, thus depriving the laboratory of its most influential supporter. The new CIW president, John C. Merriam, was showing less enthusiasm for the marine station. Until the future of the laboratory could be decided, work planned for the 1923 season was to go on, with Longley assigned to discharge "administrative responsibility." Three investigators worked at Tortugas, but the CIW sponsored the work of four others, including a report by T. Wayland Vaughan on corals collected at Samoa. In 1924 ten researchers conducted studies at Tortugas. On the last day of December 1924, the noted American biologist Thomas Hunt Morgan called a meeting in Washington, D.C., to discuss "the advantages of various laboratory sites within the region from Panama to Bermuda." Among the assembled zoologists were eight who had worked at Tortugas. Led by Longley, they spoke favorably of the laboratory on Loggerhead Key, emphasizing that "the marine flora and fauna of Tortugas are rich." They noted that the laboratory facilities were in excellent condition and downplayed "the isolation" of the station. Longley sent a memorandum of the meeting to Merriam.[4]

A year later Merriam called his own meeting "to consider the relation of the Tortugas Laboratory to other activities of the Institution in the biological sciences." Among those present were William Longley, Paul Bartsch, Charles Davenport, Herbert Jennings, and David Tennent—all supporters of the Tortugas site. Although Merriam was not ready to close the Tortugas Laboratory and believed that it had resulted in important studies, he was considering "whether at such a remote locality it would be advisable to continue" to fund the station at the same level it had thus far received. At this meeting, however, he stressed the commitment of the Carnegie Institution to supporting "its efforts in biology upon problems of growth, differentiation and evolution," which seemed to indicate that, insofar as the Tortugas Laboratory fostered such work, it would be continued.[5] At the same time, Merriam's statement suggested that the laboratory would not receive the level of support it had enjoyed under Woodward's administration and under the leadership of Mayor. Indeed, it did not, though many important contributions to marine biology were made under the auspices of the laboratory until it was finally closed for good by CIW president Vannevar Bush on August 8, 1939. Altogether, more than 140 scientists conducted research at the laboratory during the thirty-five years of its existence. In thirty-five volumes of Tortugas Laboratory papers, in separate CIW publications, and in the CIW yearbooks, scientists published in excess of two hundred papers resulting from their research at Tortugas, and many of the papers added significantly to the advancement of marine science.[6] Particularly noteworthy in this regard were truly groundbreaking work on corals by T. Wayland Vaughan and seminal studies in jellyfish taxonomy and physiology by Mayor.

To Alfred Mayor belongs the credit for conceiving the idea of the laboratory at Tortugas, for campaigning ardently to get it established, for building it, for leading its work, and for making many important scientific contributions of his own. Perhaps the words of his son Brantz six decades later sum it up as well as

anyone could: "Every stick, stone and bush; every part of Tortugas came from
his [Alfred Goldsborough Mayor's] imagination, inspiration, will and sweat. All
this started with nothing but sand and water."[7] Among Alfred Mayor's more
than one hundred publications were major contributions to hydrozoan, scyphozoan, and ctenophoran zoology and the ecology of coral reefs. Indeed, Mayor's
studies of reefs and his analysis of the effects of temperature on corals constitute
pioneering work in the field. Moreover, his *Medusae of the World* is one of the
great works on the biology of these mostly planktonic invertebrates.[8] If Mayor
originally overplayed the Dry Tortugas as an ideal site for a marine laboratory,
he nevertheless succeeded in promoting significant work there and in attracting
a notable number of able scientists to conduct research at the facility. Although
he did not get the laboratory moved to a more accessible place, Mayor was able
to broaden its mission by making expeditions to the South Pacific to conduct
comparative and ecological studies of coral reefs. His contemporaries recognized
his talent, his leadership abilities, and his devotion to the development of
marine biology.

   Of the taxa established by Mayor and Mayor with Agassiz, sixty-six hydromedusae, twelve scyphomedusae, and nine ctenophores are recognized today.
Of the taxa he established with Agassiz, nineteen hydromedusae, three scyphomedusae, and four ctenophores are currently valid; the other forty-seven taxa
of hydromedusae, nine taxa of scyphomedusae, and five taxa of ctenophores are
solely his. Altogether, then, the names of Mayor and Mayor with Agassiz are
presently associated with eighty-seven valid taxa of medusae and ctenophores.
Along with his monumental *Medusae of the World,* his superb *Ctenophores of the
Atlantic Coast of North America,* his studies of jellyfish physiology and coral-reef
ecology, and his notable direction of the Tortugas Laboratory for nearly nineteen years, these contributions assure an enduring place for him in the annals of
marine biology. Mayor's name has been affixed to at least one genus and eighteen species (see appendix 3). Among these are two scleractinian corals described
by T. Wayland Vaughan: *Coeloseris mayeri* Vaughan, 1918, and *Porites mayeri*
Vaughan, 1918. The malacologist Paul Bartsch named the land snails *Cerion
mayori* and *Opisthosiphon mayori* in honor of his longtime friend. Another
authority on corals, John W. Wells, paid tribute to Mayor in his taxon *Manicina
mayori* Wells, 1936, and the noted hydrozoan zoologist Paul Kramp honored
him in 1959 by naming a leptothecate *Melicertissa mayeri,* indicating that the
contributions of Mayor continue to enjoy recognition long after his death. Even
seven decades after he passed away, Mayor was still being honored in cnidarian
biology. The patronym *Ectopleura mayeri,* established in 1990, bears witness to
his impact on science. More recently two biologists have informally applied the
generic name *Agmayeria* to a ctenophore, and although the name has not as yet

been properly established nomenclaturally, it signifies ongoing recognition of Alfred Goldsborough Mayor as an accomplished marine biologist.

Furthermore, his contributions to the world of science are evident from the frequency of citations to his work. A survey of the "Web of Science," which provides data on scientific publications that have been cited in some 8,500 journals since 1945,[9] indicates that as of July 2005 Mayor's published works had been cited in 596 articles, an average of almost ten times per year over the sixty-year period. By far the most frequently cited of Mayor's works is the three-volume monograph *Medusae of the World,* which is listed as a reference in 176 scientific publications. Mayor's 1914 paper on the effects of temperature on tropical marine animals[10] has been cited in 79 reports, while *Ctenophores of the Atlantic Coast of North America* has been cited in 55 reports. Two of Mayor's papers on rhythmical pulsation in medusae, one in 1906 and the other in 1908,[11] have been cited 43 and 52 times respectively, and the series of three Mayor articles on Samoan corals and coral reefs, all published in 1924,[12] has been cited a combined 50 times since 1945. Added to a considerable, but unknown, number of citations prior to 1945, these references attest to the quality and lasting impact of the scientific investigations conducted by Mayor over a span of three decades. Notably, some of Mayor's works remain relevant to current research, having been cited 163 times over the past ten years (1995–2005), with 96 of those citations occurring since the year 2000.[13]

A final measure of the lasting worth of Mayor's work was the reprinting, some fifty-five years after he died, of his *Medusae of the World.* The classic monograph was published once again, in 1977, by A. Asher and Company BV, of Amsterdam, The Netherlands.

As modern marine biologists and biological oceanographers continue to advance our knowledge of life in the seas and oceans of our planet, they are building from a base established in the distant past. Indeed, tides of new information are sweeping constantly onto the shores of our minds, and we marvel over these discoveries. Yet, much remains to be explored and learned—for example, about biodiversity, about ecological changes over long periods of time, and about evolutionary relationships. As biologists continue to probe into these matters and to discover more and more about marine life, they are continuing to erect an edifice of knowledge that has its foundation in the insightful work done by their forerunners. Among those pioneers must be counted Alfred Goldsborough Mayor, a scientist of the seas.

*Appendix II*

## 1. Valid Taxa of Coelenterates Established by Alexander Agassiz and Alfred Goldsborough Mayer (Mayor)*

### *Phylum Cnidaria*

### Class Hydrozoa

| ORDER | ORIGINAL NAME AND DATE | CURRENT NAME |
|---|---|---|
| Limnomedusae | *Gonionemus suvaensis*, 1899 | *Scolionema suvaense* |
| Narcomedusae | *Solmissus marshalli*, 1902 | *Solmissus marshalli* |
| Anthoathecata | *Bougainvillea fulva*, 1899 | *Bougainvillia fulva* |
| | *Oceania pacifica*, 1899 | *Clytia pacifica* |
| | *Cytaeis vulgaris*, 1899 | *Cytaeis vulgaris* |
| | *Pandea violacea*, 1899 | *Merga violacea* |
| | *Turris pelagica*, 1902 | *Neoturris pelagica* |
| | *Pennaria vitrea*, 1899 | *Pennaria vitrea* |
| | *Lymnorea ocellata*, 1902 | *Podocoryna ocellata* |
| Leptothecata | *Oceania ambigua*, 1899 | *Clytia ambigua* |
| | *Epenthesis rangiroae*, 1902 | *Clytia rangiroae* |
| | *Dipleurosoma pacifica*, 1902 | *Dipleurosoma pacificum* |
| | *Phortis elliceana*, 1902 | *Eirene elliceana* |
| | *Eirene kambara*, 1899 | *Eirene kambara* |
| | *Eutimeta levuka*, 1899 | *Eutima levuka* |
| | *Laodicea fijiana*, 1899 | *Laodicea fijiana* |
| | *Laodicea marama*, 1899 | *Laodicea marama* |
| | *Mitrocoma mbengha*, 1899 | *Phialucium mbengha* |
| | *Staurodiscus nigricans*, 1899 | *Staurodiscus nigricans* |

*This listing is alphabetical by the current name of genus and species.

**Class Scyphozoa**

| | | |
|---|---|---|
| Rhizostomeae | *Cassiopea ndrosia*, 1899 | *Cassiopea ndrosia* |
| | *Cephea dumokuroa*, 1899 | *Netrostoma dumokuroa* |
| Coronatae | *Nausithoe picta*, 1902 | *Nausithoe picta* |

*Phylum Ctenophora*

**Class Tentaculata**

| | | |
|---|---|---|
| Cydippida | *Pleurobrachia ochracea*, 1902 | *Hormiphora ochracea* |
| | *Lampea fusiformis* | *Lampetia fusiformis* |
| Lobata | *Eucharis grandiformis*, 1899 | *Leucothea grandiformis* |

**Class Nuda**

| | | |
|---|---|---|
| Beroida | *Beroe australis*, 1899 | *Beroe australis* |

Sources of original descriptions:

Agassiz, Alexander, and Alfred Goldsborough Mayer, "Acalephs from the Fiji Islands," *Bulletin of the Museum of Comparative Zoölogy, at Harvard College* 33, no. 9 (1899): 157–89, plates 1–17.

Agassiz, Alexander, and Alfred Goldsborough Mayer, "Reports of the Scientific Results of the Expedition to the Tropical Pacific, in Charge of Alexander Agassiz, by the U.S. Fish Commission Steamer 'Albatross,' from August, 1899 to March, 1900, Commander Jefferson F. Moser, U.S.N., Commanding. III. Medusae," *Memoirs of the Museum of Comparative Zoölogy, at Harvard College* 26, no. 3 (1902): 139–76, plates 1–14.

## 2. Valid Taxa of Coelenterates Established by Alfred Goldsborough Mayer (Mayor)

*Phylum Cnidaria*

**Class Hydrozoa**

| ORDER ORIGINAL | NAME AND DATE | CURRENT NAME |
|---|---|---|
| Limnomedusae | *Cubaia*, 1894 | *Cubaia* |
| | *Cubaia aphrodite*, 1894 | *Cubaia aphrodite* |
| | *Willsia brooksi*, 1910 | *Proboscidactyla brooksi* (may be *P. stellata*) |
| Anthoathecata | *Stomotoca australis*, 1900b | *Amphinema australis* |
| | *Stomotoca rugosa*, 1900a | *Amphinema rugosum* |
| | *Dissonema turrida*, 1900b | *Amphinema turrida* |
| | *Bougainvillia frondosa*, 1900b | *Bougainvillia frondosa* |
| | *Bougainvillia niobe*, 1894 | *Bougainvillia niobe* |
| | *Tiara superba*, 1900b | *Cirrhitiara superba* |

| ORDER | ORIGINAL NAME AND DATE | CURRENT NAME |
|---|---|---|
| | *Syndictyon angulatum*, 1900a | *Coryne angulata* |
| | *Dicodonium floridana*, 1910 | *Dicodonium floridanum* |
| | *Dinema jeffersoni*, 1900b | *Dicodonium jeffersoni* |
| | *Ectopleura minerva*, 1900b | *Ectopleura minerva* |
| | *Lizzia elegans*, 1900b | *Koellikerina elegans* |
| | *Cytaeis gracilis*, 1900b | *Lizzia gracilis* |
| | *Niobia*, 1900b | *Niobia* |
| | *Niobia dendrotentacula*, 1900b | *Niobia dendrotentaculata* |
| | *Parvanemus degeneratus*, 1904 | *Pachycordyle degenerata* |
| | *Lymnorea alexandri*, 1904 | *Podocoryna alexandri* |
| | *Lymnorea borealis*, 1900a | *Podocoryna borealis* |
| | *Dysmorphosa dubia*, 1900b | *Podocoryna dubia* |
| | *Dysmorphosa minuta*, 1900b | *Podocoryna minuta* |
| | *Sarsia hargitti*, 1910 | *Sarsia hargitti* |
| | *Hybocodon forbesii*, 1894 | *Vannuccia forbesii* |
| | *Gemmaria dichotoma*, 1900b | *Zancleopsis dichotoma* |
| Leptothecata | *Blackfordia*, 1910 | *Blackfordia* |
| | *Blackfordia manhattensis*, 1910 | *Blackfordia manhattensis* |
| | *Blackfordia virginica*, 1910 | *Blackfordia virginica* |
| | *Oceania discoida*, 1900b | *Clytia discoida* |
| | *Oceania gelatinosa*, 1900b | *Clytia gelatinosa* |
| | *Oceania globosa*, 1900b | *Clytia globosa* |
| | *Oceania singularis*, 1900a | *Clytia singularis* |
| | *Tetraconnota collapsa*, 1900b | *Dipleurosoma collapsum* |
| | *Dipleurosoma ochracea*, 1910 | *Dipleurosoma ochraceum* |
| | *Phortis lactea*, 1900b | *Eirene lactea* |
| | *Eucheilota paradoxica*, 1900b | *Eucheilota paradoxica* |
| | *Multioralis ovalis*, 1900b | *Gastroblasta ovale* |
| | *Laodicea neptuna*, 1900b | *Laodicea neptuna* |
| | *Oceania carolinae*, 1900a | *Malagazzia carolinae* |
| | *Netocertoides*, 1900b | *Netocertoides* |
| | *Netocertoides brachiatum*, 1900b | *Netocertoides brachiatus* |
| | *Orchistoma tentaculata*, 1900a | *Orchistomella tentaculata* |
| | *Eucopium parvigastrum*, 1900b | *Phialella parvigastra* |
| | *Pseudoclytia*, 1900 | *Pseudoclytia* |
| | *Pseudoclytia pentata*, 1900 | *Pseudoclytia pentata* |
| | *Dipleurosoma brooksi*, 1910 | *Staurodiscus brooksi* |
| | *Toxorchis kellneri*, 1910 | *Staurodiscus kellneri* |

## Class Scyphozoa

| ORDER | ORIGINAL NAME AND DATE | CURRENT NAME |
| --- | --- | --- |
| Rhizostomeae | *Catostylus purpurus,* 1910 | *Acromitoides purpurus* |
| | *Catostylus townsendi,* 1915 | *Catostylus townsendi* |
| | *Cotylorhiza pacifica,* 1915 | *Cotylorhiza pacifica* |
| | *Lychnorhiza bartschi,* 1910 | *Crambione bartschi* |
| | *Crambione cooki,* 1910 | *Crambione cooki* |
| | *Lobonema,* 1910 | *Lobonema* |
| | *Lobonema smithi,* 1910 | *Lobonema smithi* |
| | *Phyllorhiza luzoni,* 1915 | *Phyllorhiza luzoni* |
| Semaeostomeae | *Discomedusa philippina,* 1910 | *Discomedusa philippina* |

## *Phylum Ctenophora*

## Class Tentaculata

| | | |
| --- | --- | --- |
| Cydippida | *Tinerfe beehleri,* 1912 | *Haeckelia beehleri* |
| | *Tinerfe lactea,* 1912 | *Lampea lactea* |
| | *Pleurobrachia brunnea,* 1912 | *Pleurobrachia brunnea* |
| Lobata | *Leucothea ochracea,* 1912 | *Leucothea ochracea* |
| | *Mnemiopsis mccradyi,* 1900a | *Mnemiopsis mccradyi* |

Sources of original descriptions:

Mayer, Alfred Goldsborough, "An Account of Some Medusae Obtained in the Bahamas," *Bulletin of the Museum of Comparative Zoölogy, at Harvard College* 25, no. 11 (1894): 235–42, plates 1–3.

Mayer, Alfred Goldsborough, *Ctenophores of the Atlantic Coast of North America,* Car. Inst. Wash. Publication 162 (Washington, DC: Carnegie Institution of Washington, 1912), fifty-eight pages and seventeen plates.

Mayer, Alfred Goldsborough, "Descriptions of New and Little-known Medusae from the Western Atlantic," *Bulletin of the Museum of Comparative Zoölogy, at Harvard College* 37, no. 1 (1900a): 1–27, plates 1–2.

Mayer, Alfred Goldsborough, "Medusae of the Bahamas," *Memoirs of Natural Sciences* 1, no. 1 (1904): 1–33, plates 1–7.

Mayer, Alfred Goldsborough, "Medusae of the Philippines: Being a Report upon the Scyphomedusae Collected by the United States Fisheries Bureau Steamer 'Albatross' in the Philippine Islands and Malay Archipelago, 1907–1910, and upon the Medusae Collected by the Expedition of the Carnegie Institution of Washington to Torres Straits, Australia, in 1913," *Papers of the Tortugas Laboratory* 8 (1915): 157–202, plates 1–13.

Mayer, Alfred Goldsborough, *Medusae of the World,* Car. Inst. Wash. Publication 109 (Washington, D.C.: Carnegie Institution of Washington, 1910), vols. 1 & 2, *Hydromedusae,* 1–498, plates 1–55; vol. 3, *Scyphomedusae,* 499–735, plates 56–76.

Mayer, Alfred Goldsborough, "Some Medusae from the Tortugas, Florida," *Bulletin of the Museum of Comparative Zoölogy, at Harvard College* 37, no. 2 (1900b): 13–82, plates 1–44.

## 3. Alfred Goldsborough Mayer (Mayor) Patronyms

### Generic Names
*Mayeria* Verrill, 1900 (Polychaeta)
*Agmayeria* (Ctenophora) (authorship and nomenclatural status currently unresolved)

### Specific Names
*Sertularia mayeri* Nutting, 1904 (Hydrozoa)
*Aplopus mayeri* Caudell, 1905 (Insecta)
*Cladonema mayeri* Perkins, 1906 (Hydrozoa)
*Polycitor mayeri* Hartmeyer, 1911 (Ascidiacea)
*Lobonema mayeri* Light, 1914 (Scyphozoa)
*Eutiara mayeri* Bigelow, 1918 (Hydrozoa)
*Porites mayeri* Vaughan, 1918 (Anthozoa)
*Coeloseris mayeri* Vaughan, 1918 (Anthozoa)
*Cadulus mayori* Henderson, 1920 (Mollusca)
*Cerion mayori* Bartsch, 1922 (Mollusca)
*Calcarina mayori* Cushman, 1924 (Foraminiferida)
*Hydnophora mayori* Hoffmeister, 1925 (Anthozoa)
*Rissoina mayori* Dall, 1927 (Mollusca)
*Manicina mayori* Wells, 1936 (Anthozoa)
*Thalamoporella mayori* Osburn, 1940 (Bryozoa)
*Opisthosoma mayori* Bartsch, 1946 (Mollusca)
*Melicertissa mayeri* Kramp, 1959 (Hydrozoa)
*Ectopleura mayeri* Petersen, 1990 (Hydrozoa)

## 4. Researchers and Their Topics at the Tortugas Laboratory, 1904–22

Ball, Stanley C. Yale University and later Massachusetts Agricultural College. 1913, 1914, 1917. Insects and insect migrations.

Bartsch, Paul. U.S. National Museum. 1912–17, 1919, 1921, 1922. Environmental effects on and breeding of snails of the genus *Cerion*.

Bowman, H. H. M. University of Pennsylvania. 1915, 1916. Biology of mangroves.

Brooks, William Keith. Johns Hopkins University. 1905, 1906. Biology of Ascidiacea and Decapoda; embryology of the lung in a pulmonate snail.

Cary, Lewis R. Princeton University. 1910–18; 1920. Regeneration, growth, development, and physiology of Anthozoa and in *Cassiopea;* reef structure.

Chamberlin, Rollin T. University of Chicago. 1920. Geology of Tutuila and its reefs.

Chapman, Frank Michler. American Museum of Natural History. 1907. Frigate birds and boobies of Cay Verde, Bahamas.

Clark, Hubert Lyman. Museum of Comparative Zoölogy, Harvard University. 1912–14, 1916, 1917. Studies on Echinodermata at Jamaica and in the Torres Straits.

Cole, Leon Jacob. Harvard University. 1906. Habits of ants and salps.

Conklin, Edwin Grant. Princeton University. 1905, 1907, 1915. Egg development in medusae, larval stages of Actinia.

Cowles, Rheinart Parker. Johns Hopkins University. 1905, 1906, 1908, 1909. Ecology and behavior of ghost crabs, sea stars, and brittle stars.

Cushman, Joseph A. Boston Society of Natural History. 1912, 1919. Foraminifera of Jamaica.

Dahlgren, Ulric. Princeton University. 1906, 1911, 1912, 1914–22. Electric organs of fishes; light organs of insects and marine animals.

Daly, Reginald Aldworth. Harvard University. 1919. Origin of beach rock; Pacific island geology.

Dole, R. B. U.S. Geological Survey. 1913. Chemistry of seawater at Tortugas.

Donaldson, Henry H. Wistar Institute of Anatomy. 1916. Climatic effects on rats.

Drew, George Harold. Cambridge University. 1911, 1912. Bacteria in seawater; precipitation of $CaCO_3$ in the ocean.

Drew, Gilman A. Marine Biological Laboratory, Woods Hole, Massachusetts. 1912. Cephalopod behavior and anatomy.

Edmondson, Charles H. University of Iowa. 1906. Protozoa of Tortugas.

Field, Richard M. No affiliation given. 1919. Limestones of reefs.

Gerould, John H. Dartmouth College. 1915, 1921–22. Heredity in lepidopterans.

Goldfarb, A. J. City College of New York. 1912, 1913, 1915, 1916. Research on regeneration and fertilization.

Gray, George M. Marine Biological Laboratory, Woods Hole. 1912. Collecting and faunistics, Montego Bay, Jamaica.

Gudger, Eugene W. North Carolina State Normal College. 1912–15. Anatomy of fishes.

Gulliver, F. P. Washington, D.C. 1914. Coastal geology.

Hartmeyer, Robert. Berlin Zoological Museum. 1907. Biology of ascidians and echinoderms.

Harvey, Edmund Newton. Princeton University; also at Columbia University and University of Pennsylvania. 1909–11, 1913–18, 1920–22. Animal physiology; bioluminescence.

Hatai, Shinkishi. Wistar Institute of Anatomy. 1916. Starving in *Cassiopea;* brain weights of fishes.

Hooker, Davenport. Yale University. 1905, 1907, 1908. Behavior of hatchling loggerhead sea turtles.

Ingen, Gilbert van. Princeton University. 1915. Heavy metals in invertebrates.

Jackson, Robert Tracy. Boston Society of Natural History. 1912. Biology of echinoids.

Jacobs, Merkel Henry. University of Pennsylvania. 1911. Physiology of protozoan parasites.

Jennings, Herbert Spencer. Johns Hopkins University. 1905. Behavior in sea anemones.

Jordan, Harvey Ernest. University of Virginia. 1907, 1912, 1914. Cellular biology and anatomy of invertebrates and vertebrates.

Jörgensen, E. Bergen University. 1910. Peridinea of the Tortugas.

Kellner, Carl. Johns Hopkins University. 1905–7. Embryology of appendicularians.

Lashley, Karl Spencer. Johns Hopkins University. 1913. Behavior of noddy and sooty terns.

Linton, Edwin. Washington and Jefferson College. 1906–8. Inventory of parasites.

Lipman, Charles B. University of California. 1920, 1922. Bacteria and precipitation of $CaCO_3$.

Longley, William Harding. Goucher College. 1910, 1911, 1913–22. Color and ecology of reef fishes.

Mast, S. O. Goucher College. 1910. Behavior of Turbellaria and sea turtles.

Matthai, George. Cambridge University. 1915. Biology of corals.

Mayor, Alfred Goldsborough. Carnegie Institution of Washington. 1904–22. Biology of Cnidaria and Ctenophora; studies on the Atlantic palolo worm; ecology of coral reefs; alkalinity of seawater.

McClendon, Jesse Francis. University of Missouri and later Cornell Medical College. 1908, 1910, 1916, 1917, 1919. Functional morphology, behavior, and cellular biology of invertebrates.

Meek, Seth Eugene. Field Museum of Natural History, Chicago, Illinois. 1909. Reef fishes and their habitats.

Morgulis, Sergius. University of Nebraska. 1922. Blood studies in spiny lobsters and sharks.

Mortensen, Theodor. University of Copenhagen. 1916. Development of echinoderms.

Osburn, Raymond Carroll. Columbia University. 1908. Bryozoa and Entoprocta of the Tortugas.

Perkins, H. F. University of Vermont. 1905. Biology of hydromedusae and scyphomedusae at Tortugas.

Phillips, Alexander H. Princeton University. 1915. Heavy metals in invertebrates.

Potts, Frank A. Cambridge University. 1913, 1920, 1922. Ecology of Crustacea, Torres Straits.

Pratt, Henry Sherring. Haverford College. 1909–10. Trematodes and cestodes of the Tortugas.

Reighard, Jacob Ellsworth. University of Michigan. 1905, 1907. Behavior of reef fishes.

Reinke, Edwin E. Rice Institute and later Princeton University and Vanderbilt University. 1911–14. Biology of molluscan spermatozoa.

Schaeffer, Asa A. No affiliation given. 1919, 1921, 1922. Amoebas from the Tortugas.

Setchell, William Albert. University of California. 1920, 1922. Marine and terrestrial plants of Samoa and Tahiti.

Shaw, Eugene Wesley. U.S. Geological Survey. 1915. Physiography of limestone deposits.

Silvester, Charles F. Princeton University. 1915. Studies of fishes in Puerto Rico.

Stockard, Charles R. Cornell Medical College. 1907–9. Invertebrate regeneration, behavior, and histology.

Stromsten, Frank Albert. University of Iowa. 1907, 1910. Anatomy and embryology of sea turtles.

Tashiro, Shiro. University of Chicago. 1914, 1915. Animal physiology; marine chemistry.

Tennent, David Hilt. Bryn Mawr College. 1909, 1910, 1912–13. Hybridization in echinoderms.

Tower, William L. University of Chicago. 1908–10. Heredity and variation in beetles.

Treadwell, Aaron Louis. Vassar College. 1909–10, 1913–16, 1918, 1920–21. Biology of polychaetes.

Vaughan, Thomas Wayland. U.S. Geological Survey. 1908–15, 1922. Geology; corals and coral reefs; oceanography.

Wallace, W. Seward. University of Nevada. 1908. Hydroids of the Tortugas.

Waller, J. C. Cambridge University. 1915. Electrical responses of plants.

Watson, John Broadus. University of Chicago and later Johns Hopkins University. 1907, 1910, 1912–13. Behavior of seabirds.

Wells, R. C. No affiliation given. 1919. $CO_2$ content of seawater at Tortugas.

Werber, E. I. Yale University. 1915. Development in fishes.

Zeleny, Charles. University of Indiana. 1906. Regeneration in crustaceans and *Cassiopea*.

## Abbreviations for Selected Sources

AGM Papers/SU     Alfred Goldsborough Mayer (Mayor) Papers,
Department of Special Collections, E. S. Bird Library,
Syracuse University

AMM Papers/SU     Alfred Marshall Mayer Papers,
Department of Special Collections, E. S. Bird Library,
Syracuse University

CBD Papers/APS     Charles Benedict Davenport Papers,
American Philosophical Society Library, Philadelphia,
Pennsylvania

CIW Archives     Carnegie Institution of Washington Archives,
Washington, D.C.

*CIW Year Book*     *Carnegie Institution of Washington Year Book*

H-M Papers/PU     Hyatt and Mayer (Mayor) Papers, Department of
Rare Books and Special Collections, Princeton University

HRHM Papers/SU     Harriet Randolph Hyatt Mayer (Mayor) Papers,
Department of Special Collections, E. S. Bird Library,
Syracuse University

## Chapter I—A Zeal for Zoology

1. Alfred Goldsborough Mayer, *Medusae of the World,* vol. 1: *The Hydromedusae* (Washington, D.C.: Carnegie Institution of Washington, 1910), 2.

2. Harriet Hyatt Maÿer, *The Maÿer Family* (N.p., [1911]), [4]; Alfred Goldsborough Mayer, "Autobiographical Notes," typescript dated January 1917, with addendum [1919], in Brantz and Ana Mayor Private Collection, Hanover, N.H.; Charles B. Davenport, *Biographical Memoir of Alfred Goldsborough Mayor, 1868–1922* (Washington, D.C.: National Academy of Sciences, 1926), 21:1–2; Alfred Goldsborough Mayer to Harriet Hyatt Mayer, November 13, 1912, in HRHM Papers/SU.

3. H. H. Maÿer, *Maÿer Family,* [4]; A. G. Mayer and Robert S. Woodward, *Biographical Memoir of Alfred Marshall Mayer, 1836–1897* (Washington, D.C.: National Academy of Sciences, 1916), 7:243–50; Alfred M. Mayer, "Eulogy on Joseph Henry,"

*Proceedings of the American Association for the Advancement of Science* 29 (1881): 69; M. M. Fisher and John J. Rice, *History of Westminster College, 1851–1903* (Columbia, Mo.: Press of E. W. Stephens, 1908), 25–28, 44–45; Samuel Gring Hefelbower, *The History of Gettysburg College, 1833–1932* (Gettysburg, Pa.: Gettysburg College, 1932), 466; A. G. Mayer to J. Percy Moore, April 16, 1902, in Ewell Sale Stewart Library, Academy of Natural Sciences of Philadelphia; Alfred M. Mayer, manuscript notebook, "Memoranda of Studies pursued in Paris, Sept.–April 1864–65," in Alfred M. Mayer Papers, Rare Books and Manuscript Division, New York Public Library; manuscript of entry on Alfred Marshall Mayer for *Appleton's Cyclopedia of American Biography,* in Marcus Benjamin Papers, Smithsonian Institution Archives, Washington, D.C.; Alfred M. Mayer to Lewis Mayer, April 24, 1861, in Manuscript Division, Maryland Historical Society Library, Baltimore, Maryland.

4. A. G. Mayer and Woodward, *Biographical Memoir,* 7:250–51; A. G. Mayer to J. Percy Moore, April 16, 1902, in Academy of Natural Sciences of Philadelphia; A. M. Mayer to F. A. P. Barnard, April 24, 1872, in H-M Papers/PU; A. G. Mayer, "Autobiographical Notes"; Davenport, *Biographical Memoir,* 21:1.

5. A. G. Mayer, "Autobiographical Notes"; Davenport, *Biographical Memoir,* 21:1–2; Alfred G. Mayer, "The Radiation and Absorption of Heat by Leaves," *American Journal of Science* 45 (1893): 340–46; Lucian I. Blake to A. M. Mayer, March 7, 1892, in AMM Papers/SU.

6. A. M. Mayer to Alpheus Hyatt, February 19, 1892, in AMM Papers/PU; Alfred G. Mayer, "On an Improved Heliostat Invented by Alfred M. Mayer," *American Journal of Science,* 4th ser., 4 (1897): 306–8; Alfred G. Mayer, "Alpheus Hyatt, 1838–1902," *Popular Science Monthly* 78 (1911): 129–46; A. M. Mayer to Alexander Agassiz, January 24, 1874, in H-M Papers/PU; Ralph W. Dexter, "The Annisquam Sea-side Laboratory of Alpheus Hyatt, Predecessor of the Marine Biological Laboratory at Woods Hole, 1880–1886," in *Oceanography: The Past,* ed. M. Sears and D. Merriam (New York: Springer-Verlag, 1980), 94–100; Mary P. Winsor, *Reading the Shape of Nature: Comparative Zoology at the Agassiz Museum* (Chicago: University of Chicago Press, 1991), 135–39.

7. Alfred G. Mayer, "Habits of the Box Tortoise," *Popular Science Monthly* 38 (November 1890): 60–65; Alfred G. Mayer, "Habits of the Garter Snake," *Popular Science Monthly* 42 (February 1893): 485–88; Davenport, *Biographical Memoir,* 21:4; A. G. Mayer, "Autobiographical Notes."

8. A. G. Mayer, "Autobiographical Notes"; Davenport, *Biographical Memoir,* 21:1; Garland E. Allen, "Davenport, Charles Benedict," in *American National Biography* (New York: Oxford University Press, 1999), 6:126–28; Alfred G. Mayer, "On the Color and Color-Patterns of Moths and Butterflies," *Bulletin of the Museum of Comparative Zoölogy, at Harvard College* 30 (1897): 169; Alfred G. Mayer, "Effects of Natural Selection and Race-Tendency upon the Color-Patterns of Lepidoptera," *Science Bulletin* 1 (1902): 31; A. G. Mayer to Charles B. Davenport, August 2, 1895, in CBD Papers/APS.

9. Mayor kept two detailed journals of the voyage, dated January 4–March 7, 1893. These are in the AGM Papers/SU.

10. A. G. Mayer, "Autobiographical Notes"; A. G. Mayer, *Medusae of the World,* 1:1. The illustration, dated "March 12, 1893," is located in the Alfred M. Mayer Papers, Rare Books and Manuscript Division, New York Public Library.

11. Alfred G. Mayer, "An Account of Some Medusae Obtained in the Bahamas," *Bulletin of the Museum of Comparative Zoölogy, at Harvard College* 25 (1894): 235–42, three plates with twelve figures.

12. A. G. Mayer, "Autobiographical Notes"; A. G. Mayer to Davenport, September 17 and August 2, 1895, in CBD Papers/APS. Some of Mayor's zoology class notebooks are in the Mayer Papers, Rare Books and Manuscript Division, New York Public Library. One labeled "Zool. VI" and another titled "Geology" contain notes on butter-flies. Four other Mayor class notebooks are in the private collection of Michael B. Mayor, M.D., Hanover, N.H.. Mayor's trip to Europe in 1894 is documented in the *New York Times,* June 16, 1894. On the back of a photograph of himself, taken in a French studio, Mayor noted that he was there in 1891. In a letter dated April 5, 1891, he said he was declining his parents' invitation to accompany them to Europe, but it appears that he went anyway. The letter is in the AMM Papers/SU, and the photograph is in the Brantz and Ana Mayor Private Collection, Hanover, N.H..

13. Unidentified newspaper clipping [1905], in CBD Papers/APS; A. G. Mayer to H. Hyatt, August 27, 1897, in HRHM Papers/SU; Davenport, *Biographical Memoir,* 21:1.

14. Alfred G. Mayer, " The Development of the Wing Scales and Their Pigment in Butterflies and Moths," *Bulletin of the Museum of Comparative Zoölogy, at Harvard College* 29 (1896): 209–36, and seven plates of seventy-four figures; A. G. Mayer, "On the Color and Color-Patterns of Moths and Butterflies," ibid., 30 (1897): 169–229, and four plates of sixty figures.

15. Alfred G. Mayer, "Journal of a Journey around the World," manuscript in AGM Papers/SU.

16. Ibid.; Alexander Agassiz and Alfred Goldsborough Mayer, "On Some Medusae from Australia," *Bulletin of the Museum of Comparative Zoölogy, at Harvard College* 32 (1898): 15–19, and three plates.

17. A. G. Mayer, "Journal of a Journey"; Davenport, *Biographical Memoir,* 21:2; Agassiz to A. G. Mayer, October 6, 1896, in Alexander Agassiz Letterbook, Museum of Comparative Zoology Library, Cambridge, Massachusetts; Davenport to A. G. Mayer, October 30, 1896, in H-M Papers/PU. *The Historical Record of Harvard University, 1636–1936* (Cambridge: Harvard University, 1937) indicates that Mayor was "Assistant in Charge of Radiates (Museum of Comparative Zoology) 1894–1900," but the letter from Agassiz on October 6, 1896, refers to a salary for Mayor beginning that year. The official record of Mayor's degree of S.D., in 1897, in "natural history" is in *Quinquennial Catalogue of the Officers and Graduates of Harvard University* (Cambridge: Harvard University, 1905), 520.

18. Alfred G. Mayer, "A New Hypothesis of Seasonal Dimorphism in Lepidoptera," *Psyche* 8 (1897): 47–50, 59–62; Ronald L. Numbers, *Darwinism Comes to America* (Cambridge: Harvard University Press, 1998), 33–40.

19. Alfred G. Mayer, "Alexander Agassiz, 1835–1910," *Popular Science Monthly* 77 (1910): 422–23; Alfred G. Mayer, *Sea-shore Life: The Invertebrates of the New York Coast and the Adjacent Coast Region* (New York: New York Zoological Society, 1905), 24; Winsor, *Reading the Shape of Nature,* 147–62; Edward Lurie, "Agassiz, Alexander," in *American National Biography,* 24 vols. (New York: Oxford University Press, 1999), 1:172–73.

20. A. G. Mayer to H. Hyatt, June 1, 1897, in HRHM Papers/SU; Alfred G. Mayer, "Some Medusae from the Tortugas, Florida," *Bulletin of the Museum of Comparative Zoölogy, at Harvard College* 37 (1900): 42, 51.

21. A. G. Mayer to H. Hyatt, June 20, 1897, in HRHM Papers/SU; John H. Davis, Jr., "The Ecology of the Vegetation and Topography of the Sand Keys of Florida," *Papers from Tortugas Laboratory* 33 (November 1942): 176.

22. A. G. Mayer to H. Hyatt, [March 1897] and July 17 and 19, 1897, in HRHM Papers/SU; A. G. Mayer to Davenport, [ca. July 20, 1897], in CBD Papers/APS; Julia Ward Howe to A. G. Mayer, July 21, 1897, in H-M Papers/PU; *New York Times,* July 14, 1897; *New York Tribune,* July 14, 1897.

23. A. G. Mayer to H. Hyatt, August 15, 17, and 20, 1897, in HRHM Papers/SU; A. G. Mayer to Davenport, in CBD Papers/APS.

24. A. G. Mayer to H. Hyatt, August 22, 28, and 31, 1897, in HRHM Papers/SU.

25. Lester D. Stephens and Dale R. Calder, "John McCrady of South Carolina: Pioneer Student of North American Hydrozoa," *Archives of Natural History* 19 (1992): 39–54; A. G. Mayer to H. Hyatt, [September 8] and 9, 1897, in HRHM Papers/SU; Edward McCrady to A. G. Mayer, September 28, 1897, in AGM Papers/SU.

26. A. G. Mayer to H. Hyatt, September 6, 8, and 9, 1897, in HRHM Papers/SU.

27. A. G. Mayer to H. Hyatt, September 10, 1897, in HRHM Papers/SU; H. Hyatt to A. G. Mayer, September 8, 1897, and October 8 [1897], in AGM Papers/SU. On Agassiz's relationship to his assistants, see Winsor, *Reading the Shape of Nature,* 207–21.

28. A. G. Mayer to H. Hyatt, October 18, 1897 [two letters, one of which Mayer did not complete until October 27, 1897], in HRHM Papers/SU.

29. A. G. Mayer to H. Hyatt, October 18–27, 1897, in HRHM Papers/SU.

30. A. G. Mayer to H. Hyatt, October 28 [completed ca. November 9], 1897, in HRHM Papers/SU.

31. Ibid.; A. G. Mayer to H. Hyatt, December 15, 1897, in HRHM Papers/SU; Alfred G. Mayer, portion of manuscript journal labeled "The Fiji Islands, November 7–28, 1897," in AGM Papers/SU. Another portion of the same manuscript journal, dated "Dec. 2–Dec. 9, 1897," is in the Brantz and Ana Mayor Private Collection, Hanover, N.H.

32. A. G. Mayer to H. Hyatt, December 24, 1897, and January 1 [misdated 1897] and 2, 1898, in HRHM Papers/SU; Alfred G. Mayer, manuscript journal ["The Fiji Islands"], "Nov. 29 [1897]–Jan. 13 [1898]," in AGM Papers/SU; portion of same journal, "Dec. 2–Dec. 9, 1897," in Brantz and Ana Mayor Private Collection, Hanover, N.H.

33. A. G. Mayer to H. Hyatt, January 22, 1898, and February 10, 14, and 27, 1898, in HRHM Papers/SU.

34. A. G. Mayer to H. Hyatt, February 27 and 28, and March 3, 1898, in HRHM Papers/SU.

35. Alexander Agassiz and Alfred Goldsborough Mayer, "On *Dactylometra,*" *Bulletin of the Museum of Comparative Zoölogy, at Harvard College* 32 (1898): 1–11, and thirteen plates.

36. A. G. Mayer to H. Hyatt, June 13 and July 5 and 17, 1898, in HRHM Papers/SU; H. Hyatt to A. G. Mayer, July 10 and 15, 1898, in AGM Papers/SU.

37. Winsor, *Reading the Shape of Nature,* 222–24; A. G. Mayer to H. Hyatt, July 5 and 17, 1898, in HRHM Papers/SU; A. G. Mayer to Davenport, August 3, 1898, in CBD Papers/APS.

38. A. G. Mayer to H. Hyatt, August 19 and September 19, 1898, in HRHM Papers/SU; A. G. Mayer to Davenport, September 22, 1898, in CBD Papers/APS; A. G. Mayer, "Autobiographical Notes."

39. H. Hyatt to A. G. Mayer, August 29, 1898, in HRHM Papers/SU.

40. A. G. Mayer to H. Hyatt, two undated letters but both in September 1898, in HRHM Papers/SU; Alexander Agassiz and Alfred Goldsborough Mayer, "Acalephs from the Fiji Islands," *Bulletin of the Museum of Comparative Zoölogy, at Harvard College* 32 (1899): 157–89, and seventeen plates.

41. See Jane Maienschein, *Transforming Traditions in American Biology, 1880–1915* (Baltimore: Johns Hopkins University Press, 1991), esp. 13–69; and Sally Gregory Kohlstedt, "Museums on Campus: A Tradition of Inquiry and Teaching," Keith R. Benson, "From Museum Research to Laboratory Research," and Toby Appel, "Organizing Biology: The American Society of Naturalists and Its 'Affiliated Societies,' 1883–1923"—all in *The American Development of Biology,* ed. Ronald Rainger, Keith R. Benson, and Jane Maienschein (Philadelphia: University of Pennsylvania Press, 1988), 15–47, 49–83, and 87–120, respectively.

42. A. G. Mayer to Agassiz, May 5, 1899, in H-M Papers/PU; Ernst Haeckel, *Das System der Medusen: Erster Theil einer Monographie der Medusen* (Jena: Verlag von Gustav Fischer, 1879), 360 pp.

43. A. G. Mayer to Agassiz, May 5, 1899, in H-M Papers/PU.

44. A. G. Mayer to Davenport, May 13, 1899, in CBD Papers/APS.

45. A. G. Mayer to Agassiz, May 14, 1899, in HRHM Papers/SU; H. Hyatt to A. G. Mayer, June 12, 1899, in AGM Papers/SU; H. Hyatt to A. G. Mayer, May 28, 1899, in HRHM Papers/SU.

46. A. G. Mayer to H. Hyatt, May 19 and June 23 and 26, 1899, in HRHM Papers/SU; A. G. Mayer to H. S. Jennings, June 25, 1899, in Herbert Spencer Jennings Papers, American Philosophical Society Library, Philadelphia, Pennsylvania; A. G. Mayer to Agassiz, July 5, 1899, in H-M Papers/PU.

47. A. G. Mayer to H. Hyatt, June 23, [1899], in HRHM Papers/SU.

48. H. Hyatt to A. G. Mayer, July 6, 1899, in AGM Papers/SU; A. G. Mayer to H. Hyatt, July 29, 1899, in HRHM Papers/SU.

49. A. G. Mayer to Jennings, August 23, 1899, in Herbert Spencer Jennings Papers, American Philosophical Society Library; A. G. Mayer, manuscript "Notebook 6," in Mayer Papers, Rare Books and Manuscript Division, New York Public Library; Alfred Goldsborough Mayor, "Pacific Island Reveries," typescript, ca. 1920, in AGM Papers/SU; Alexander Agassiz, "Preliminary Report and List of Stations [of the Expedition to the Tropical Pacific, August 1899–March 1900]," *Memoirs of the Museum of Comparative Zoölogy, at Harvard College* 26 (1902): 1–114; Alfred G. Mayer, "Journal of the Cruise of the U.S. F[ish] C[ommission] S[teamer] 'Albatross' in the South Pacific from August 23, 1899, [to] March 4, 1900," in Brantz and Ana Mayor Private Collection, Hanover, N.H.

50. Alfred Goldsborough Mayer, "Some Species of *Partula* from Tahiti: A Study in Variation," *Memoirs of the Museum of Comparative Zoölogy, at Harvard College* 26 (1902): 117–35, and one plate; A. G. Mayer, manuscript "Notebook 7, September 29–November 15, 1899," in Mayer Papers, Rare Books and Manuscript Division, New York Public Library; A. G. Mayer, "Journal of the Cruise of the . . . 'Albatross.'"

51. Agassiz, "Preliminary Report," 143; A. G. Mayer, "Journal of the Cruise of the . . . 'Albatross.'"

52. Agassiz, "Preliminary Report " (see "Track of the Albatross"); A. G. Mayer to Davenport, December 18, 1899, in CBD Papers/APS; A. G. Mayer to H. Hyatt, March 4, 1900, in HRHM Papers/SU; A. G. Mayer, "Notebook 7"; A. G. Mayer, "Pacific Island Reveries"; Maple Club of Japan, March 6, 1900, to A. G. Mayer [translation of invitation to a dinner], in H-M Papers/PU; A. G. Mayer, "Journal of the Cruise of the . . . 'Albatross.'"

53. Alfred Goldsborough Mayer, "On the Mating Instinct in Moths," *Annals and Magazine of Natural History,* 7th ser., 5 (1900): 183–90; also published in *Psyche* 9 (February 1900): 15–20. Soon thereafter Mayer presented a paper at the Marine Biological Laboratory, "On the Development of Color in Moths and Butterflies," *Biological Lectures from the Marine Biological Laboratory, Tenth Lecture* (1900): 157–64.

54. A. G. Mayer, "On the Mating Instinct in Moths."

55. A. G. Mayer to H. Hyatt, two undated letters [April 1900], in HRHM Papers/SU; A. G. Mayer to Mrs. Charles B. Davenport, April 18, 1900, in CBD Papers/APS; Franklin W. Hooper to A. G. Mayer, April 18 and June 15, 1900, in H-M Papers/PU.

56. Agassiz to A. G. Mayer, June 30, 1900, in Alexander Agassiz Letterbook, Museum of Comparative Zoology Library, Cambridge, Massachusetts.

57. Alfred Goldsborough Mayer, "Descriptions of New and Little-known Medusae from the Western Atlantic," *Bulletin of the Museum of Comparative Zoölogy, at Harvard College* 37 (June 1900): 1–9, and six plates.

58. A. G. Mayer, "Some Medusae from the Tortugas,"13–82, and forty-five plates.

### Chapter II—Campaigning Curator

1. Lee Richard Hiltzik, "The Brooklyn Institute of Arts and Sciences' Biological Laboratory, 1890–1924: A History" (Ph.D. diss., State University of New York at Stonybrook, 1993), 5, 23–36, 53–54, 122–42, 177–86; Rebecca Hooper Eastman, *The Story of the Brooklyn Institute of Arts and Sciences, 1824–1924* (N.p., [1925]), 3–13; Brooklyn Institute of Arts and Sciences, *Thirteenth Year Book of the Brooklyn Institute of Arts and Sciences* (Brooklyn: The Institute, 1901), title page and 268–70; "Hooper, Franklin William," *National Cyclopaedia of American Biography,* 63 vols. to date (New York: J. T. White, 1891– ), 13:46–47.

2. A. G. Mayer to Hooper, November 22 and 28, 1900, in Alfred G. Mayer Correspondence, Brooklyn Museum Archives.

3. A. G. Mayer to Hooper, December 1 and 4, 1900, in ibid.; Bashford Dean to A. G. Mayer, October 5, 1901, in H-M Papers/PU.

4. A. G. Mayer to Hooper, December 10, 1900, in Brooklyn Museum Archives; Brooklyn Institute of Arts and Sciences, *Thirteenth Year Book,* 269–70; Brooklyn Institute of Arts and Sciences, *Prospectus for 1900–1901* (Brooklyn: The Institute, 1900),

190–93. Other lectures arranged or presented by Mayer are listed in the Brooklyn Institute's yearbooks for 1902–4 and in each prospectus of the institute for 1902 and 1903.

5. A. G. Mayer to Hooper, January 21, 1901, and Hooper to A. G. Mayer, January 22, 1901, in Alfred G. Mayer Correspondence, Brooklyn Museum Archives.

6. Hooper to A. G. Mayer, January 22, 1901, in ibid.

7. A. G. Mayer to Hooper, March 4, 1901, in ibid.

8. A. G. Mayer to Hooper, April 23, 1901, in ibid.

9. A. G. Mayer to Hooper, May 30, 1901, and Hooper to A. G. Mayer, May 31, 1901, both in ibid.

10. A. G. Mayer to H. Hyatt, [June 12, 1901], in HRHM Papers/SU.

11. A. G. Mayer to Davenport, January 8 and 31, March 10, and May 8, 1901, in CBD Papers/APS.

12. Alfred Goldsborough Mayer, "The Variations of a Newly-Arisen Species of Medusa," *Science Bulletin* 1 (April 1901): 1–27, and two plates; Dale R. Calder, *Shallow-Water Hydroids of Bermuda: The Thecatae, Exclusive of the Plumularioidea,* Royal Ontario Museum Life Sciences Contributions, 154 (Toronto: Royal Ontario Museum, 1990), 65, 68.

13. Alexander Agassiz and Alfred Goldsborough Mayer, "Medusae [of the Tropical Pacific]," *Memoirs of the Museum of Comparative Zoölogy, at Harvard College* 26 (1902): 139–75, and fourteen plates.

14. Alfred G. Mayer to "Trustees of the Carnegie Institution, Washington, D.C.," January 2, 1902, and notation on same, dated December 18, 1902, in AGM Papers/SU; A. G. Mayer to H. Hyatt, February 13, [February, n.d.], and June 9, 1902, in HRHM Papers/SU.

15. James Trefil and Margaret Hindle Hazen, *Good Seeing: A Century of Science at the Carnegie Institution of Washington, 1902–2002* (Washington, D.C.: Joseph Henry Press, 2002), 21–35; James D. Ebert, "Carnegie Institution of Washington and Marine Biology: Naples, Woods Hole, and Tortugas," *Biological Bulletin* 168 (June 1985): 172–73.

16. Howard S. Miller, *Dollars for Research: Science and Its Patrons in Nineteenth-Century America* (Seattle: University of Washington Press, 1970), 166–81; Jane Maienschein, "Introduction," in *Centennial History of the Carnegie Institution of Washington,* vol. 5: *The Department of Embryology,* ed. Jane Maienschein, Marie Glitz, and Garland E. Allen (Cambridge: Cambridge University Press, 2004), 2–4.

17. Charles A. Kofoid, "The Biological Stations of Europe," *United States Bureau of Education Bulletin, 1910,* no. 4 (1910): 1; C. M. Yonge, "Development of Marine Biological Laboratories," *Science Progress* 64 (January 1956): 1–15.

18. Keith R. Benson, "Laboratories on the New England Shore: The 'Somewhat Different Direction' of American Marine Biology," *New England Quarterly* 61 (March 1988): 55–78; Edward Lurie, *Louis Agassiz: A Life in Science* (Chicago: University of Chicago Press, 1960), 380–81; Winsor, *Reading the Shape of Nature,* 198–207.

19. Benson, "Laboratories," 55–78; Dexter, "Annisquam Sea-side Laboratory," 94–100; Jane Maienschein, "Agassiz, Hyatt, Whitman, and the Birth of the Marine Biological Laboratory," *Biological Bulletin* 6 (1985): 26–34.

20. Jane Maienschein, *One-Hundred Years Exploring Life, 1888–1988: The Marine Biological Laboratory at Woods Hole* (Boston: Jones and Bartlett Publishers, 1989),

19–26; Frank R. Lillie, *The Woods Hole Marine Biological Laboratory* (Chicago: University of Chicago Press, 1944), 47–61.

21. Alfred G. Mayer to "The President of the Carnegie Institution," April 7, 1902, in Alfred G. Mayer Correspondence, Brooklyn Museum Archives; A. G. Mayer to Davenport, [1910], in CBD Papers/APS.

22. Alfred G. Mayer to "The President of the Carnegie Institution," April 7, 1902, in Alfred G. Mayer Correspondence, Brooklyn Museum Archives.

23. A. G. Mayer to Hooper, April 9, 16, and 23, 1902, in Alfred G. Mayer Correspondence, Brooklyn Museum Archives.

24. A. G. Mayer to Agassiz, April 28, 1902, and to Hooper, May 2 and 7, 1902, in Alfred G. Mayer Correspondence, Brooklyn Museum Archives.

25. Agassiz to A. G. Mayer, May 22, 1902, and A. G. Mayer to Hooper, May 23, 1902, in Alfred G. Mayer Correspondence, Brooklyn Museum Archives; A. G. Mayer to Harriet Hyatt Mayer, May 31, 1902, in HRHM Papers/SU; *CIW Year Book, 1902,* 1 (1903): 271–74 (a letter from Agassiz to "Dr. Billings").

26. Ebert, "Carnegie Institution," 174–80; Lillie, *Woods Hole Marine Biological Laboratory,* 52–56; Maienschein, *One-Hundred Years Exploring Life,* 19–26.

27. A. G. Mayer to H. Hyatt Mayer, June 23 and July 7, 1902, in HRHM Papers/SU; A. G. Mayer, "Effects of Natural Selection," 31–86, and two plates.

28. A. G. Mayer to H. Hyatt Mayer, September 19, 1902, in HRHM Papers/SU; Alfred Goldsborough Mayer, "Medusae of the Hawaiian Islands Collected by the Steamer *Albatross* in 1902," *Bulletin of the United States Fish Commission for 1903* 23, pt. 3 (1906): 1133–43, and three plates.

29. A. G. Mayer to Davenport, June 3, 1902, and Hooper to A. G. Mayer, June 3, July 5, and August 2, 1902, in Alfred G. Mayer Correspondence, Brooklyn Museum Archives; Alfred Goldsborough Mayer, "The Tortugas, Florida, as a Station for Research in Biology," *Science,* n.s. 17 (January 30, 1903): 190–92. On relevant discussions of the rise and development of marine biological laboratories, see Yonge, "Development of Marine Biological Laboratories," 1–15; Benson, "Laboratories," 55–78; Keith R. Benson, "Review Paper: The Naples Stazione Zoologica and Its Impact on the Emergence of American Marine Biology," *Journal of the History of Biology* 21 (Summer 1988): 331–41.

30. A. G. Mayer, "Tortugas, Florida," 190–92; Maienschein, *Transforming Traditions in American Biology,* 96–104.

31. A. G. Mayer to M. A. Bigelow, January 16, 1903, in Alfred G. Mayer Correspondence, Brooklyn Museum Archives. Favorable letters include: Maynard M. Metcalf to A. G. Mayer, February 10, 1903; William H. Dall to A. G. Mayer, February 10, 1903; Frank M. Chapman to A. G. Mayer, [February 1903]; George H. Parker to A. G. Mayer, February 11, 1903; Charles S. Minot to A. G. Mayer, February 13, 1903; Dean to A. G. Mayer, February 14, 1903; and J. Playfair McMurrich to A. G. Mayer, February 16, 1903—all in H-M Papers/PU. Unfavorable letters include N. V. Neal to A. G. Mayer, February 24, 1903; and Edward L. Mark to A. G. Mayer, May 21, 1903—both in H-M Papers/PU.

32. Thomas H. Montgomery to A. G. Mayer, February 11, 1903; Frank R. Lillie to A. G. Mayer, February 14, 1903; Charles C. Nutting to A. G. Mayer, February 16 and

24, 1903; and David Starr Jordan to A. G. Mayer, February 17, 1903—all in H-M Papers/PU.

33. A. G. Mayer to Davenport, February 18, 1903, and A. G. Mayer to Agassiz, March 6, 1903, in Alfred G. Mayer Correspondence, Brooklyn Museum Archives; Agassiz to A. G. Mayer, March 6, 1903, and August 11, [1903], and Davenport to A. G. Mayer, February 11 and 26, 1903, in H-M Papers/PU; A. G. Mayer to Davenport, February 14, March 2, and May 7 and 25, 1903, in CBD Papers/APS.

34. T. H. Morgan to A. G. Mayer, February 12, 1903, E. G. Conklin to A. G. Mayer, February 23, 1903, and C. O. Whitman to A. G. Mayer, May 2, 1903, in H-M Papers/ PU.

35. A. G. Mayer to Davenport, March 2, 1903, in Alfred G. Mayer Correspondence, Brooklyn Museum Archives; A. G. Mayer to Whitman, May 6, 1903, in H-M Papers/ PU.

36. Alfred Goldsborough Mayer, "A Tropical Marine Laboratory for Research?," *Science,* n.s. 17 (April 24, 1903): 655–60.

37. Davenport to A. G. Mayer, copy of letter to *Science* [n.d.], in Alfred G. Mayer Correspondence, Brooklyn Museum Archives; Charles B. Davenport, "The Proposed Biological Station at the Tortugas," *Science,* n.s., 17 (June 12, 1903): 945–47. Other favorable letters on the proposed laboratory include ones from Charles C. Nutting and William E. Ritter to *Science,* n.s. 17 (May 22, 1903): 823–25 and 825, respectively. Unfavorable letters include ones from Hubert Lyman Clark, *Science,* n.s., 17 (June 19, 1903): 979–80; and P. H. Rolf, *Science,* n.s., 17 (June 26, 1903): 1008–9.

38. Alfred Goldsborough Mayer, "The Status of Public Museums in the United States," *Science,* n.s., 17 (May 29, 1903): 843–51; Richard Rathbun to A. G. Mayer, July 29, 1903, and A. G. Mayer to Rathbun, August 25, 1903, in Record Unit 55, Smithsonian Institution Archives. For more on Davenport, see Garland E. Allen, "Heredity, Development, and Evolution," in *Centennial History of the Carnegie Institution,* vol. 5: *The Department of Embryology,* ed. Jane Maienschein, Marie Glitz, and Garland E. Allen (Cambridge: Cambridge University Press, 2004), 145–56.

39. A. G. Mayer to H. H. Mayer, October 26, 1903, in HRHM Papers/SU. In her genealogy of the family, Harriet Mayor spelled her first daughter's name "Katherine" (see H. H. Maÿer, *Maÿer Family*), but it appears later as "Katharine."

40. Alfred Goldsborough Mayer, "Medusae of the Bahamas," *Memoirs of Natural Science, The Museum of the Brooklyn Institute of Arts and Sciences Memoirs of Natural Science* 1 (April 1904): 1–33, and seven plates.

41. A. G. Mayer to H. H. Mayer, October 21 and 26, 1903, in HRHM Papers/SU.

### Chapter III—A Speck in the Sea

1. L. Wayne Landrum, *Fort Jefferson and the Dry Tortugas National Park* (Big Pine Key, Fla.: the author, 2003), 1–72.

2. Neil E. Hurley, *Lighthouses of the Dry Tortugas, An Illustrated History* (Aiea, Hawaii: Historic Lighthouse Publishers, 1994), 1–80.

3. Ibid., 23–28.

4. Landrum, *Fort Jefferson,* 39; Hurley, *Lighthouses,* 24.

5. John James Audubon, quoted in Hurley, *Lighthouses,* 13.

6. Charles Cleveland Nutting, "American Hydroids, Part II: Sertularidae," in *Smithsonian Institution, United States National Museum Special Bulletin,* no. 4 (Washington, D.C.: Smithsonian Institution, 1904), 58.

7. A. August Healey to A. G. Mayer, February 12, 1904, Jennings to A. G. Mayer, February 16, 1904, and A. G. Mayer to Jennings, February 29, 1904, in H-M Papers/ PU; Hooper to A. G. Mayer, February 13, 1904, and Charles D. Walcott to A. G. Mayer, January 25, 1904, in AGM Papers/SU; A. G. Mayer to Davenport, February 16, 1904, in CBD Papers/APS; Ebert, "Carnegie Institution," 180; Trefil and Hazen, *Good Seeing,* 50; *CIW Year Book, 1904,* 3 (1905): 22; "The Search for the 'Exceptional Man,'" *New York Times,* March 5, 1905, SM1; Scott Derks, ed., *The Value of a Dollar: Prices and Incomes in the United States, 1860–1999* (Lakeville, Conn.: Grey House, 1999), 2.

8. J. H. Davis, "Ecology of the Vegetation and Topography," 178–84; [Alfred G. Mayer], "Detailed Plans of Work, Equipment and Expenses of the Carnegie Marine Laboratory for Research at the Tortugas, Florida," typescript, January 28, 1904, in Tortugas Laboratory Papers, American Philosophical Society Library.

9. A. G. Mayer, "Detailed Plans."

10. A. G. Mayer to Davenport, March 28, April 14, and May 19, 1904, in CBD Papers/APS; *CIW Year Book, 1905,* 4 (1906): 108–13.

11. A. G. Mayer, "Detailed Plans"; *CIW Year Book, 1904,* 3 (1905): 50–55; A. G. Mayer to H. H. Mayer, June 22 and 29, and July 12, 1904, in HRHM Papers/SU.

12. *CIW Year Book, 1904,* 3 (1905): 50–54.

13. Agassiz to A. G. Mayer, August 4 and 11, 1904, in Alexander Agassiz Letterbook, Museum of Comparative Zoology Library, Cambridge, Massachusetts; Agassiz to A. G. Mayer, August 6, 1904, in H-M Papers/PU; *CIW Year Book, 1905,* 4 (1906): 108–10.

14. A. G. Mayer to H. H. Mayer, September 27, 1904, in HRHM Papers/SU; A. G. Mayer, *Sea-shore Life,* 9, 15–16, 24.

15. T.B., Review of A. G. Mayer's *Sea-shore Life, American Naturalist* 40 (May 1906): 378; C[harles] H. T[ownsend], "Alfred Goldsborough Mayor," *New York Zoological Society Bulletin* 25 (November 1922): 138; Madison Grant to A. G. Mayer, June 26, 1905, and A. G. Mayer to H. H. Mayer, June 21, 1906, in HRHM Papers/SU.

16. A. G. Mayer to Walcott, December 15, 1904, A. G. Mayer to Robert S. Woodward, February 6, 1905, and A. G. Mayer to Walter M. Gilbert, April 2, 1905, in CIW Archives; W. K. Brooks to A. G. Mayer, February 24, [1905], in H-M Papers/PU; *CIW Year Book, 1905,* 4 (1906): 113–24.

17. *CIW Year Book, 1905,* 4 (1906): 113–24; A. G. Mayer to Woodward, May 16 and [ca. June 1], 1905, in CIW Archives; A. G. Mayer to H. H. Mayer, May 2 and 26, 1905, in HRHM Papers/SU.

18. A. G. Mayer to H. H. Mayer, April 11, June 27 and 29, and August 1, 1905, in HRHM Papers/SU. Conard survived the illness, completed the Ph.D. degree in 1906, and became a highly productive scholar, serving as a professor in Grinnell College for most of his career.

19. *CIW Year Book, 1905,* 4 (1906): 113.

20. A. G. Mayer to Woodward, [n.d., but date of receipt indicated as June 6, 1905], in CIW Archives.

21. A. G. Mayer to H. H. Mayer, May 26 and June 12 and 16, 1905, in HRHM Papers/SU; A. G. Mayer to Woodward, [ca. June 1, 1905], in CIW Archives.

22. L. M. Passano, "Scyphozoa and Cubozoa," in *Electrical Conduction and Behaviour in "Simple" Invertebrates*, ed. G. A. B. Shelton (Oxford: Clarendon Press, 1982), 149–202.

23. Alfred Goldsborough Mayer, "Rhythmical Pulsation in Scyphomedusae," *Carnegie Institution of Washington Publication*, no. 47 (1906): 1–62.

24. Alfred G. Mayer and Caroline G. Soule, "Some Reactions of Caterpillars and Moths," *Journal of Experimental Zoology* 3 (1906): 415–33.

25. P. H. Rolfs, letter to editor "The Proposed Biological Laboratory at the Tortugas," *Science*, n.s., 17 (June 26, 1903): 1009; Alfred G. Mayer, "Tortugas Marine Laboratory of the Carnegie Institution of Washington: Its Aims and Its Problems," lecture presented at the American Museum of Natural History, noted in the *New York Times*, February 18, 1906; A. G. Mayer to Davenport, March 24, 1905, in CBD Papers/APS; Thomas H. Montgomery, Jr., to A. G. Mayer, February 8, 1906, in H-M Papers/PU; A. G. Mayer to H. H. Mayer, May 30, 1906, in HRHM Papers/SU.

26. A. G. Mayer to H. H. Mayer, March 17, 23, and 27, 1907, in HRHM Papers/SU; Winsor, *Reading the Shape of Nature*, 223.

27. A. G. Mayer to H. H. Mayer, April 23 and 28, 1906, in HRHM Papers/SU.

28. A. G. Mayer to H. H. Mayer, March 23, April 3, and June 13, 1906, in HRHM Papers/SU.

29. A. G. Mayer to Davenport, November 11, 1906, in CBD Papers/APS; Jane Maienschein, "First Impressions: American Biologists at Naples," *Biological Bulletin* 168, supplement (June 1985): 191.

30. A. G. Mayer to Woodward, February 4, 1907, in CIW Archives; A. G. Mayer to H. H. Mayer, March 5 and March 30–April 11, 1907, in HRHM Papers/SU; *CIW Year Book, 1907*, 6 (1908): 107; George Shiras III, "One Season's Game-Bag with the Camera," *National Geographic Magazine* 19 (June 1908): 388–402.

31. A. G. Mayer to H. H. Mayer, March 30–April 11, 1907, in HRHM Papers/SU; Shiras, "One Season's Game-Bag," *National Geographic Magazine* 19 (June 1908): 388–402.

32. *CIW Year Book, 1907*, 6 (1908): 108–23; Kerry W. Buckley, *Mechanical Man: John Broadus Watson and the Beginnings of Behaviorism* (New York: Guilford Press, 1989), 200, note 33.

33. A. G. Mayer to Woodward, April 28, 1907, in CIW Archives; *CIW Year Book, 1907*, 6 (1908): 106–8, 118–19; John B. Watson to A. G. Mayer, August 8, [1907], and September 26, 1907, and Conklin to A. G. Mayer, September 19, 1907, in H-M Papers/PU; A. G. Mayer to H. H. Mayer, April 23, May 3 and 30, and June 21 and 28, 1907, in HRHM Papers/SU. Two of the papers Mayor presented at the International Zoological Congress were published: "The Annual Swarming of the Atlantic Palolo," *Proceedings of the Seventh International Zoological Congress* [August 19–24, 1906] (1912): 147–51; and "The Cause of Rhythmical Pulsation in Scyphomedusae," ibid., 278–81. Listed by title only in the same volume is A. G. Mayer, "Vantage Grounds for the Study of the Marine Life of the West Indian Region," 964. The fourth paper was not located in the volume.

34. A. G. Mayer to H. H. Mayer, August 2 and 26, 1907, in HRHM Papers/SU. In his "Presidential Address," Alexander Agassiz called for a comparative study of marine fauna on the eastern and western coasts of Panama, but he apparently opposed the motion to study the possibly adverse effects of the Panama Canal on marine life (mentioned in the summary of the "General Meeting"), in *Proceedings of the Seventh International Zoological Congress* (1912): 55–59 and 969, respectively.

35. Indeed, Anna Hyatt became famous as a sculptor, with many of her works now on exhibit at Brookgreen Gardens in coastal South Carolina. Brookgreen Gardens, the first sculpture garden in the United States, was founded in the 1930s by Anna and her philanthropist husband Archer Huntington. Several of Harriet Mayor's sculptures are also on display there.

36. H. H. Mayer to A. G. Mayer, June 30 and August 7, 1907, in AGM Papers/SU; A. G. Mayer to Davenport, November 25, 1907, in CBD Papers/APS; A. G. Mayer to H. H. Mayer, [ca. December 1907 or January 1908], in HRHM Papers/SU; *CIW Year Book, 1908,* 7 (1908): 225; Brantz Mayor, "Alfred Goldsborough Mayor," in *The Alfred G. Mayor and Katharine M. Townsend Memorial Fund* (Woods Hole, Mass.: Woods Hole Oceanographic Institution, 1985), 36. (Note: Katharine Mayor Townsend was the second child of Alfred G. Mayor. Her mother originally spelled the name "Katherine.")

37. Alfred Goldsborough Mayer, "Our Neglected Southern Coast: A Cruise of the Carnegie Institution Yacht 'Physalia,'" *National Geographic Magazine* 19 (1908): 859–71.

38. W. L. Tower to A. G. Mayer, April 8 and May 15, 1908, in H-M Papers/PU; *CIW Year Book, 1908,* 7 (1908): 29, 118–38.

39. *CIW Year Book, 1908,* 7 (1908): 119.

40. Ibid.

41. A. G. Mayer to H. H. Mayer, April 15, 1908, in HRHM Papers/SU.

42. A. G. Mayer to H. H. Mayer, May 10, 1908, in HRHM Papers/SU.

43. Alfred Goldsborough Mayer, "Should Our Colleges Establish Summer Schools?," *Science,* n.s., 23 (May 4, 1906): 703–4; Alfred Goldsborough Mayer, "Marine Laboratories, and Our Atlantic Coast," *American Naturalist* 42 (1908): 533–36; Alfred Goldsborough Mayer, "Autonomy for the University?," *Science,* n.s. 30 (November 12, 1909): 673–75; A. G. Mayer to H. H. Mayer, July 11, 1908, in HRHM Papers/SU; T. Wayland Vaughan to A. G. Mayer, August 12, 1908, in H–M Papers/PU.

44. *CIW Year Book, 1909,* 8 (1910): 28, 125–28; A. G. Mayer to Woodward, April 21, 1909, in CIW Archives; A. G. Mayer to H. H. Mayer, April 23 and May 9, 1909, in HRHM Papers/SU.

45. A. G. Mayer to H. H. Mayer, April 12, 1909, in HRHM Papers/SU.

46. *CIW Year Book, 1909,* 8 (1910): 125–53.

47. Ibid., 127–28; *Papers from the Tortugas Laboratory of the Carnegie Institution of Washington,* vols. 1 and 2 (1907).

48. *Papers from the Tortugas Laboratory of the Carnegie Institution of Washington,* vol. 2.

49. Alfred Goldsborough Mayer, "The Research Work of the Tortugas Laboratory," *Popular Science Monthly* 76 (April 1910): 397–411, and twelve photographs.

50. Ibid., 401–11.

51. H. H. Mayer to A. G. Mayer, April 2, [1910], in AGM Papers/SU; A. G. Mayer, "Alexander Agassiz," 419–46.

52. A. G. Mayer, "Alexander Agassiz," 422–23, 435, 439–42, 446.

53. Henry B. Fine to A. G. Mayer, June 29, 1911, and Conklin to A. G. Mayer, October 20 and November 16, 1911, in H-M Papers/PU; H. H. Mayer to A. G. Mayer, April 14, May 24, and June 22, 1911, in AGM Papers/SU; A. G. Mayer to Davenport, November 5, 1910, in CBD Papers/APS; "Harriet Hyatt Mayer," *Brookgreen Bulletin* (Summer 1973): n.p.; *New York Times,* October 9, 1910.

54. *CIW Year Book, 1910,* 9 (1911): 117–20; *CIW Year Book, 1911,* 10 (1912): 22; A. G. Mayer to H. H. Mayer, June 7 and July 8, 1910, in HRHM Papers/SU.

55. A. G. Mayer to H. H. Mayer, June 2 and July 12, 1910, in HRHM Papers/SU.

56. Woodward to Walcott, June 18, 1910, in Charles Doolittle Walcott Collection, Smithsonian Institution Archives, Washington, D.C.

57. *CIW Year Book, 1910,* 9 (1911): 117–20.

## Chapter IV—Monographs on Medusae

1. A. G. Mayer, *Medusae of the World,* vol. 1: *The Hydromedusae,* 1–230; vol. 2: *The Hydromedusae,* 231–498; vol. 3: *The Scyphomedusae,* 499–735, *Carnegie Institution of Washington Publication,* no. 109 (Washington, D.C.: Carnegie Institution of Washington, 1910).

2. Ibid., 1:1.

3. Ibid.

4. Ibid., 1:2.

5. Ibid., 1:4.

6. Ibid., 1:2.

7. The date of publication of each volume of *Medusae of the World* is stamped overleaf of the title page in copies examined at libraries of the Royal Ontario Museum, the University of Toronto, the Virginia Institute of Marine Science, the South Carolina Marine Resources Research Institute, and the University of Georgia.

8. A. G. Mayer, *Medusae of the World,* 3:499.

9. J. Bouillon and F. Boero, "Synopsis of the Families and Genera of the Hydromedusae of the World, with a List of the Worldwide Species," *Thalassia Salentina* 24 (2000): 47–296.

10. A. C. Marques and A. G. Collins, "Cladistic Analysis of Medusozoa and Cnidarian Evolution," *Invertebrate Biology* 123 (2004): 23–42.

11. Charles C. Nutting, Review of A. G. Mayer's *Medusae of the World,* vols. 1 and 2, *Science,* n.s., 32 (1910): 596–97.

12. Anonymous, "A Monograph of the Jellyfishes," *Nature* 85 (1910): 285–87.

13. *CIW Year Book, 1910,* 9 (1911): 123.

14. Ernst Haeckel to A. G. Mayer, April 15, 1912, in H-M Papers/PU.

15. Haeckel, *Das System der Medusen,* 1–360; Ernst Haeckel, *System der Acraspeden: Zweite Hälfte des Systems der Medusen* (Jena: Verlag von Gustav Fischer, 1880), 361–672.

16. A. G. Mayer, *Medusae of the World,* 1:2.

17. Paul L. Kramp, "The Hydromedusae of the Atlantic Ocean and Adjacent Waters," *Dana-Report* 46 (1959): 1–283; Paul L. Kramp, "Synopsis of the Medusae of the World," *Journal of the Marine Biological Association of the United Kingdom* 40 (1961): 1–469; Paul L. Kramp, "The Hydromedusae of the Pacific and Indian Oceans," *Dana-Report*, secs. 2 and 3, 72 (1968): 1–200.

18. Kramp, "Synopsis," 8.

19. F. S. Russell, *The Medusae of the British Isles: Anthomedusae, Leptomedusae, Limnomedusae, Trachymedusae and Narcomedusae* (Cambridge: Cambridge University Press, 1953), 1–530; F. S. Russell, *The Medusae of the British Isles, II: Pelagic Scyphozoa with a Supplement to the First Volume on Hydromedusae* (Cambridge: Cambridge University Press, 1970), 1–284.

20. D. V. Naumov, "Gidroidy I gidromeduzy morskikh, solonovatovodnykh I presnovodnykh basseinov SSR," *Akademiya Nauk SSSR, Opredeliteli po Faune SSSR* 70 (1960): 1–626, translated as *Hydroids and Hydromedusae of the USSR* (Jerusalem: Israel Program for Scientific Translations, 1969), catalog no. 5108, pp. 1–660; D. V. Naumov, "Stsifoidnye meduzy morei SSSR," *Akademiya Nauk SSSR, Opredeliteli po Faune SSSR* 75 (1961): 1–98; P. F. S. Cornelius, "Donat Vladimirovitch Naumov (1921–1984)," in *Modern Trends in the Systematics, Ecology, and Evolution of Hydroids and Hydromedusae*, ed. J. Bouillon, F. Boero, F. Cicogna, and P. F. S. Cornelius (Oxford: Oxford University Press, 1987), 5–7.

21. B. W. Halstead, *Poisonous and Venomous Marine Animals of the World*, 2nd rev. ed. (Princeton, N.J.: Darwin Press, 1988), 1–1168; J. A. Williamson, P. F. Fenner, J. W. Burnett, and J. F. Rifkin, eds., *Venomous and Poisonous Marine Animals: A Medical and Biological Handbook* (Sydney: University of New South Wales Press, 1996), 1–504.

22. Mary N. Arai, *A Functional Biology of Scyphozoa* (London: Chapman and Hall, 1997), 68–91, 203–12; Jennifer E. Purcell and Mary N. Arai, "Interactions of Pelagic Cnidarians and Ctenophores with Fish: A Review," *Hydrobiologia* 451 (2001): 27–44.

23. A. G. Mayer, *Medusae of the World*, 2:500–504.

24. Kramp, "Hydromedusae of the Atlantic," 75; Kramp, "Synopsis," 9; Kramp, "Hydromedusae of the Pacific," 4–5; Russell, *Medusae of the British Isles*, 21; Russell, *Medusae of the British Isles, II*, 1–10.

25. A. G. Mayer, *Medusae of the World*, 3:570–71, 578.

26. Ibid., 1:3.

27. Ibid.

28. Ibid., 2:344.

29. Ibid., 3:622; C. F. W. Krukenberg, "Über den Wassergehalt der Medusen," *Zoologischer Anzeiger* 3 (1880): 306.

30. Arai, *Functional Biology*, 216–18; Russell, *Medusae of the British Isles, II*, 85–86.

31. Mary N. Arai and Anita Brinckmann-Voss, "Hydromedusae of British Columbia and Puget Sound," *Canadian Bulletin of Fisheries and Aquatic Sciences* 204 (1980): 117.

32. A. G. Mayer, *Medusae of the World*, 2:247, 308.

33. Halstead, *Poisonous and Venomous Marine Animals*, passim; Williamson et al., *Venomous and Poisonous Marine Animals*, passim.

34. J. E. Purcell, W. M. Graham, and H. J. Dumont, "Jellyfish Blooms: Ecological and Societal Importance," *Hydrobiologia* 451 (2001): 1–222; Claudia E. Mills, "Jellyfish

Blooms: Are Populations Increasing Globally in Response to Changing Ocean Conditions?," *Hydrobiologia* 451 (2001): 55–68; W. M. Graham, D. L. Martin, D. L. Felder, V. L. Asper, and H. M. Perry, "Ecological and Economic Implications of a Tropical Jellyfish Invader in the Gulf of Mexico," *Biological Invasions* 5 (2000): 53–69.

35. M. Omori and E. Nakano, "Jellyfish Fisheries in Southeast Asia," *Hydrobiologia* 451 (2001): 19–26.

36. A. G. Mayer, *Medusae of the World,* 3:517, 537, 706.

37. Ibid., 3:499, 503, 520; Marques and Collins, "Cladistic Analysis," 23–42.

38. A. G. Mayer, *Medusae of the World,* 2:239.

39. Ibid., 2:357, 3:596.

40. Alfred Goldsborough Mayer, "Medusae of the Philippines and of Torres Strait," *Carnegie Institution of Washington Publication,* no. 212 (1915): 157–202; Alfred Goldsborough Mayer, "Report upon the Scyphomedusae Collected by the United States Bureau of Fisheries Steamer 'Albatross' in the Philippine Islands and Malay Archipelago," *United States National Museum Bulletin* 100 (1917): 175–233; *CIW Year Book, 1918,* 17 (1919): 153.

41. *CIW Year Book, 1915,* 14 (1916): 185; T. W. Schmidt and L. Pikula, *Scientific Studies on Dry Tortugas National Park: An Annotated Bibliography,* National Park Service, Current References 97-1 (Washington, D.C.: U.S. Department of Commerce, National Oceanic and Atmospheric Administration, National Oceanographic Data Center, and Department of the Interior, 1997), 1–108.

42. A. G. Mayer, *Medusae of the World,* 1:2.

## Chapter V—Remote Reefs

1. A. G. Mayer, "Alpheus Hyatt," 129–46; A. G. Mayer to H. H. Mayer, April 2 and July 2, 1911, in HRHM Papers/SU; H. H. Mayer to A. G. Mayer, July 20, 1911, in AGM Papers/SU.

2. A. G. Mayer to H. H. Mayer, February 9, 1912, in HRHM Papers/SU; H. H. Mayer to A. G. Mayer, July 15, 1912, in AGM Papers/SU; H. H. Mayer to [Mother], August 17, 18, [19?], 20, and 24, and September 9 and 12, 1911 [from Sharon, Mass., sanatorium], in Brantz and Ana Mayor Private Collection, Hanover, N.H..

3. A. G. Mayer to William E. Ritter, March 6 and September 11, 1911, in William E. Ritter Papers, Bancroft Library, University of California at Berkeley; Helen Raitt and Beatrice Moulton, *Scripps Institution of Oceanography: First Fifty Years* (N.p.: The Ward Ritchie Press, 1967), 3–44, 64–67.

4. *CIW Year Book, 1911,* 10 (1912): 120–23; A. G. Mayer to H. H. Mayer, June 6 and 24 and July 18, 1911, in HRHM Papers/SU.

5. *CIW Year Book, 1911,* 10 (1912): 122. Drew's place in studying denitrifying bacteria in sea water is mentioned in Eric L. Mills, *Biological Oceanography: An Early History, 1870–1960* (Ithaca, N.Y.: Cornell University Press, 1989), 70, 96–97.

6. *CIW Year Book, 1911,* 10 (1912): 122.

7. Alfred Goldsborough Mayer, *Ctenophores of the Atlantic Coast of North America* (Washington, D.C.: Carnegie Institution of Washington, 1912), 60 pp.; G. Richard Harbison and Laurence P. Madin, "An Appreciation of Alfred G. Mayor," in *The Alfred G.*

*Mayor and Katharine M. Townsend Memorial Fund* (Woods Hole, Mass.: Woods Hole Oceanographic Institution, 1985), 45–47.

8. Alfred Goldsborough Mayer, "An Atlantic 'Palolo,' *Staurocephalus gregaricus,*" *Bulletin of the Museum of Comparative Zoölogy, at Harvard College* 36 (1900): 1–14, and three plates; A. G. Mayer, "The Atlantic Palolo," *Science Bulletin* 1 (1902): 93–103, and one plate; Leonard B. Clark and Walter N. Hess, "Swarming of the Atlantic Palolo Worm, *Leodice fucata* (Ehlers)," *Papers from the Marine Biological Laboratory of the Carnegie Institution of Washington* 33 (1940): 23–33; Alfred Goldsborough Mayer, "The Annual Breeding-Swarm of the Atlantic Palolo," *Publication of the Carnegie Institution of Washington,* no. 102 (1908): 105–12, and one plate; A. G. Mayer to H. H. Mayer, June 23 and July 8, 1902, and August 1, 1905, in HRHM Papers/SU; A. G. Mayer to Davenport, July 7, 1902, in CBD Papers/APS.

9. A. G. Mayer, "Annual Swarming of the Atlantic Palolo," 147–51; A. G. Mayer, "Annual Breeding-Swarm of the Atlantic Palolo," 105–12.

10. *CIW Year Book, 1912,* 11 (1913): 119–29; A. G. Mayer to H. H. Mayer, December 11, 1911, February 14 and 17, March 24, April 11 and 24, and May 25, 1912, in HRHM Papers/SU; Hubert Lyman Clark to A. G. Mayer, November 25 and December 21, 1911, and April 5, 1912, in H-M Papers/PU; A. G. Mayer to Davenport, April 2, 1912, in CBD Papers/APS.

11. A. G. Mayer to Woodward, August 31 and September 30, 1912, in CIW Archives; G. Harold Drew to A. G. Mayer, August 10, October 8 and 30, and November 18, 1912, in H-M Papers/PU.

12. A. G. Mayer to Woodward, August 22, 1912, and "Project of T. Wayland Vaughan," [August 1912], in CIW Archives.

13. A. G. Mayer to H. H. Mayer, May 26–28, and July 16, 1912, in HRHM Papers/SU; H. H. Mayer to A. G. Mayer, June 19, July 10 and 29, and October 31, 1912, in AGM Papers/SU. Additional information on the stay in the sanatorium is in Harriet Mayor's notebook of letters, drawings, and other items, in the Brantz and Ana Mayor Private Collection, Hanover, N.H.

14. Woodward to A. G. Mayer, August 20, 1912, and A. G. Mayer to Woodward, December 17 and 21, 1912, in CIW Archives; A. G. Mayer to H. H. Mayer, December 25, 1912, in HRHM Papers/SU.

15. A. G. Mayer to H. H. Mayer, February 12, 1913, in HRHM Papers/SU; H. H. Mayer to A. G. Mayer, February 14, 1913, in AGM Papers/SU; A. G. Mayer to Woodward, February 11, 1913, in CIW Archives.

16. A. G. Mayer to H. H. Mayer, January 20, February 14, and March 26, 1913, in HRHM Papers/SU.

17. Vaughan to A. G. Mayer, January 20, 1913, in CIW Archives; Vaughan to A. G. Mayer, January 26, [1913], in H-M Papers/PU.

18. Vaughan to A. G. Mayer, January 20, 1913, in CIW Archives; Frederic Wood Jones, "On the Growth-forms and Supposed Species in Corals," *Proceedings of the Zoological Society of London* 31 (June 18, 1907): 518–56; *CIW Year Book, 1908,* 7 (1909): 118; T. Wayland Vaughan, "Geology of the Florida Keys and the Marine Deposits and Recent Corals of Southern Florida," *CIW Year Book, 1908,* 7 (1909), 135.

19. A. G. Mayer to H. H. Mayer, January 26 and June 5, 1912, in HRHM Papers/ SU; A. G. Mayer to Woodward, July 5, 1913, in CIW Archives.

20. Vaughan to A. G. Mayer, July 29, 1913, in CIW Archives.

21. Ibid.

22. Ibid.

23. A. G. Mayer to Woodward, September 10, 1913, in CIW Archives.

24. *CIW Year Book, 1913,* 12 (1914): 163; William Saville-Kent, *The Great Barrier Reef of Australia: Its Products and Potentialities* (London: W. H. Allen & Co., [1893]), passim; A. G. Mayer to Woodward, September 10, 1913, in CIW Archives; Alfred Goldsborough Mayer, "An Expedition to the Coral Reefs of Torres Straits," *Popular Science Monthly* 85 (September 1914): 210; Hubert Lyman Clark, *Carnegie Scientists in the Antipodes* (Boston, 1914), 1–6.

25. A. G. Mayer, "Expedition," 212–22; A. G. Mayer to H. H. Mayer, August 31, September 4, and October 30, 1913, in HRHM Papers/SU; Clark, *Carnegie Scientists,* 5–7.

26. A. G. Mayer to H. H. Mayer, November 1, 1913, in HRHM Papers/SU.

27. A. G. Mayer, "Expedition," 21–27; Alfred Goldsborough Mayer, "Ecology of the Murray Island Coral Reef," *Proceedings of the National Academy of Sciences* 1 (1915): 211–14; Alfred Goldsborough Mayer, "Ecology of the Murray Island Coral Reef," *Papers from the Department of Marine Biology of the Carnegie Institution of Washington* 9 (1918): 3–48; T. Wayland Vaughan, "Corals and the Formation of Coral Reefs," in *Annual Report of the Smithsonian Institution, 1917* (Washington, D.C.: The Smithsonian Institution, 1919), 189–238. Developments in ecology and biological oceanography had been under way for some time, and though the field was making significant advances by this time, it came into its own during the 1930s. See E. L. Mills, *Biological Oceanography,* passim, esp. 262–83.

28. A. G. Mayer, "Ecology of the Murray Island Coral Reef," *Papers from the Department of Marine Biology,* 3–48; C. M. Yonge, "The Royal Society and the Study of Coral Reefs," in *Oceanography: The Past,* ed. M. Sears and D. Merriam (New York: Springer-Verlag, 1980), 445.

29. A. G. Mayer, "Ecology of the Murray Island Coral Reef," *Papers from the Department of Marine Biology,* 7–17.

30. A. G. Mayer, "Expedition," 231; A. G. Mayer to H. H. Mayer, December 1, 1913, and March 14, 1914, in HRHM Papers/SU; *CIW Year Book, 1913,* 12 (1914): 169–72; J. Stanley Gardiner to A. G. Mayer, March 11, 1914, in AGM Papers/SU.

31. *CIW Year Book, 1918,* 17 (1919): 164–65.

32. F. Wood Jones to A. G. Mayer, December 14, 1914, in H-M Papers/PU; *CIW Year Book, 1914,* 13 (1915): 174; A. G. Mayer to H. H. Mayer, April 18, 1914, in HRHM Papers/SU.

33. A. G. Mayer to H. H. Mayer, April 20 and 24, 1914, in HRHM Papers/SU.

34. A. G. Mayer to H. H. Mayer, May 2, 7, 12, 19, and 20, and June 25, 1914, in HRHM Papers/SU; A. G. Mayer to Woodward, June 2 and July 19, 1914, in CIW Archives.

35. H. H. Mayer to A. G. Mayer, April 2, May 17, and August 5, 1914, in AGM Papers/SU; A. G. Mayer to H. H. Mayer, August 2, 1914, in HRHM Papers/SU; H. H.

Mayer, "Notes from Rouen to A.G.M., August 1914," manuscript journal in the Brantz and Ana Mayor Private Collection, Hanover, N.H.

36. H. H. Mayer, "Notes from Rouen"; A. G. Mayer to H. H. Mayer, August 13, 23, and 24, 1914, in HRHM Papers/SU; A. G. Mayer to Frank A. Potts, August 14, 1914, and H. H. Mayer to A. G. Mayer, August 10 and 13 and September 3, 1914, in AGM Papers/SU.

37. *CIW Year Book, 1914,* 13 (1915): 177–85.

38. Ibid., 177, 185–90; *CIW Year Book, 1915,* 14 (1916): 186.

39. A. G. Mayer to H. H. Mayer, January 13, 1913, in HRHM Papers/SU; A. G. Mayer to Davenport, June 8, 1914, in CBD Papers/APS.

40. Alfred Goldsborough Mayer, ["The Mission of Scientists"], Address of the retiring vice president of section F of the American Association for the Advancement of Science, *Science* 61 (January 15, 1915): 81–82.

41. J. Stanley Gardiner to A. G. Mayer, January 8, 1915, and H. H. Mayer to A. G. Mayer, May 9, 1915, in AGM Papers/SU; Woodward to U.S. secretary of state, March 16, 1915, in CIW Archives.

42. A. G. Mayer to H. H. Mayer, April 5, May 19 and 31, and June 14, 1915, in HRHM Papers/SU; *CIW Year Book, 1915,* 14 (1916): 182–84.

43. *CIW Year Book, 1915,* 14 (1916): 187–93; A. G. Mayer, "Medusae of the Philippines and of Torres Straits," 157–203; A. G. Mayer, "Report upon the Scyphomedusae," 173–233.

44. *CIW Year Book, 1914,* 13 (1915): 190; *CIW Year Book, 1915,* 14 (1916): 183, 185.

45. Passano, "Scyphozoa and Cubozoa," 149–202.

46. L. M. Passano to D. R. Calder, September 17, 2004, in possession of Dale R. Calder, Royal Ontario Museum, Toronto, Ontario, Canada.

47. George John Romanes, "The Croonian Lecture: Preliminary Observations on the Locomotor System of Medusae," *Philosophical Transactions of the Royal Society of London* 166 (1876): 269–313; George John Romanes, "Further Observations on the Locomotor System of Medusae," *Philosophical Transactions of the Royal Society of London* 167 (1877): 659–752; George John Romanes, "Concluding Observations on the Locomotor System of Medusae," *Philosophical Transactions of the Royal Society of London* 171 (1880): 161–202.

48. Alfred Goldsborough Mayer, "Sub-marine Solution of Limestone in Relation to the Murray-Agassiz Theory of Coral Atolls," *Proceedings of the National Academy of Sciences* 2 (1916): 28–30; Gerald J. Bakus, "The Biology and Ecology of Tropical Holothurians," in *Biology and Geology of Coral Reefs,* ed. O. A. Jones and R. Endean (New York and London: Academic Press, 1973), 2:339–44.

49. *CIW Year Book, 1916,* 15 (1917): 117; A. G. Mayer to H. H. Mayer, March 19 and April 8, 1916, in HRHM Papers/SU; A. G. Mayer to Woodward, January 18 and February 6, 1916, in CIW Archives.

50. A. G. Mayer to H. H. Mayer, April 8 and May 24, 1916, in HRHM Papers/SU; *CIW Year Book, 1916,* 15 (1917): 19–20, 171–79; *CIW Minutes of the Meeting of the Executive Committee, Thursday, January 18, 1917,* 5.

51. Alfred Goldsborough Mayer, "A History of Fiji," *Popular Science Monthly* 86 (1915): 521–38, and 87 (1915): 31–49, 292–306, and *Scientific Monthly* 1 (1915): 18–35; Alfred Goldsborough Mayer, "A History of Tahiti," *Popular Science Monthly* 86 (1915): 105–27, 403–66; Alfred Goldsborough Mayer, "Papua, Where the Stone-Age Lingers," *Scientific Monthly* 1 (1915): 105–23; Alfred Goldsborough Mayer, "The Islands of the Mid-Pacific," *Scientific Monthly* 2 (1916): 125–45; Alfred Goldsborough Mayer, "Java, the Exploited Island," *Scientific Monthly* 2 (1916): 350–54; Alfred Goldsborough Mayer, "The Men of the Mid-Pacific," *Scientific Monthly* 2 (1916): 5–26; Arthur L. Day to A. G. Mayer, April 29, 1916, and Sir William Macgregor to A. G. Mayer, December 20, 1916, in H-M Papers/PU; A. G. Mayer to Davenport, July 14, 1916, in CBD Papers/APS.

52. A. G. Mayer to Davenport, May 12, October 28, and November 2, 3, and 29, 1916, in CBD Papers/APS; Alfred Goldsborough Mayer, "William Stimpson, 1832–1872," *National Academy of Sciences Biographical Memoir* 8 (1918): 419–33; Ronald S. Vasile, "Stimpson, William," *American National Biography Online;* letter from Ronald S. Vasile to Lester Stephens, September 10, 2004, in possession of Lester D. Stephens, Athens, Georgia. For additional information on Davenport and the eugenics movement, see Hamilton Cravens, *The Triumph of Evolution: American Scientists and the Heredity-Environment Controversy, 1900–1941* (Philadelphia: University of Pennsylvania Press, 1978), 15–17, 43–45, 49–51.

53. A. G. Mayer to Davenport, February 17, 1917, in CBD Papers/APS.

54. *CIW Year Book, 1917,* 16 (1918): 161; A. G. Mayer to H. H. Mayer, February 17 and 20 and March 4, 1917, in HRHM Papers/SU.

55. *CIW Year Book, 1917,* 16 (1918): 161; William M. Davis to A. G. Mayer, January 27 and February 4, 1916, in H-M Papers/PU.

56. Alfred Goldsborough Mayer, "Coral Reefs of Tutuila, with Reference to the Murray-Agassiz Solution Theory," *Proceedings of the National Academy of Sciences* 3 (1917): 522–26.

57. Alfred Goldsborough Mayer, "Observations upon the Alkalinity of the Surface Water of the Tropical Pacific," *Proceedings of the National Academy of Sciences* 3 (1917): 548–52; *CIW Year Book, 1917,* 16 (1918): 163.

## Chapter VI—Twilight at Tortugas

1. A. G. Mayer to H. H. Mayer, May 23 and June 16, 1917, in HRHM Papers/SU; Alfred Goldsborough Mayer, "Is Death from High Temperature Due to the Accumulation of Acid in the Tissues?," *Proceedings of the National Academy of Sciences* 3 (1917): 626–27; A. G. Mayer, "Is Death from High Temperature Due to the Accumulation of Acid in the Tissues?," *American Journal of Physiology* 44 (1917): 581–85; Alfred G. Mayer, "Efficacy of Holothurians in Dissolving Limestone," *CIW Year Book, 1917,* 16 (1918): 186–87; Alfred Goldsborough Mayer, "On the Non-existence of Nervous Shell-Shock in Fishes and Marine Invertebrates," *Proceedings of the National Academy of Sciences* 3 (1917): 597–98.

2. *CIW Year Book, 1917,* 16 (1918): 163; A. G. Mayer to H. H. Mayer, July 1 and August 1, 1917, in HRHM Papers/SU; A. G. Mayer to Davenport, October 27, 1917, in CBD Papers/APS.

3. A. G. Mayer to Woodward [dictated to, and written by, H. H. Mayer], December 9 and 27, 1917, and [Woodward] to H. H. Mayer, December 19, 1917, in CIW Archives.

4. A. G. Mayer to Walcott, December 23, 1917, and January 15, 1918, Walcott to A. G. Mayer, January 7, 1918, and Walcott to H. H. Mayer, January 24 and May 2, 1919, in Charles D. Walcott Collection, Record Unit 7004, Smithsonian Institution Archives, Washington, D.C.

5. A. G. Mayer to Woodward, January 15 and 18, 1918, in CIW Archives; A. G. Mayer to H. H. Mayer, February 9, 1918, in HRHM Papers/SU.

6. A. G. Mayer to Woodward, January 10 and March 22, 1918, Woodward to U.S. secretary of state, May 27, 1918, and A. G. Mayer's affidavit, May 27, 1918, in CIW Archives; *CIW Year Book, 1918,* 17 (1919): 149–50.

7. A. G. Mayer to Woodward, May 29, 1919, inspector general to A. G. Mayer, July 25, 1918, and commissioner-warden and magistrate, Tobago, October 22, 1918, in CIW Archives.

8. A. G. Mayer to H. H. Mayer, March 15, 1918, in HRHM Papers/SU; Alfred Goldsborough Mayor, *Navigation, Illustrated by Diagrams* (Philadelphia and London: J. B. Lippincott, 1918), 1–3; *The Princeton Bric-a-Brack* 44 (June 1, 1919): 99; A. G. Mayor to Jennings, November 2, 1918, in H. S. Jennings Papers, American Philosophical Society Library, Philadelphia, Pennsylvania; A. G. Mayer (Mayor), addendum and handwritten note in "Autobiographical Notes," in Brantz and Ana Mayor Private Collection, Hanover, N.H.

9. A. G. Mayer to H. H. Mayer, June 7 and July 23, 1918, in HRHM Papers/SU; *CIW Year Book, 1918,* 150.

10. Alfred Goldsborough Mayer, "The Growth-Rate of Samoan Coral Reefs," *CIW Year Book, 1915,* 14 (1916): 168–70; Alfred Goldsborough Mayor, "The Growth-Rate of Samoan Coral Reefs," *Proceedings of the National Academy of Sciences* 4 (1918): 390–93; Alfred Goldsborough Mayor, "Structure and Ecology of the Samoan Coral Reefs," *Papers from the Department of Marine Biology of the Carnegie Institution of Washington* 19 (1924): 1–25; P. L. Colin, "A Brief History of the Tortugas Marine Laboratory," in *Oceanography: The Past,* ed. M. Sears and D. Merriam (New York: Springer-Verlag, 1980), 142; A. L. Dahl and A. E. Lamberts, "Environmental Impact on a Samoan Coral Reef: A Resurvey of Mayor's 1917 Transect," *Pacific Science* 31 (1977): 309–19.

11. *CIW Year Book, 1918,* 150–51.

12. A. G. Mayor to H. H. Mayor, January 1, 1919, in HRHM Papers/SU; A. G. Mayor to Joseph A. Cushman, March 3, 1919, in Paleobiology Division, United States National Museum, Washington, D.C.; A. G. Mayor to Woodward, February 18, 1919, and Woodward to A. G. Mayor, April 9, 1919, in CIW Archives.

13. A. G. Mayor to Woodward, February 8, 1919, in CIW Archives.

14. Woodward to A. G. Mayor, February 24, 1919, and A. G. Mayor to Woodward, February [24?] and 26, and May 17, 1919, in CIW Archives; Woodward to Walcott, February 18, 1919, in Charles D. Walcott Collection, Smithsonian Institution Archives, Washington, D.C.; Alfred Goldsborough Mayor, "Detecting Ocean Currents by Observing Their Hydrogen-ion Concentration," *Proceedings of the American Philosophical Society* 58 (1919): 150–60.

15. A. G. Mayor to H. H. Mayor, May 25 and June 3, 1919, and numerous letters from the Mayor children to their father, in HRHM Papers/SU; "A Hyatt Mayor, Former Curator of Prints at the Metropolitan, 78," *New York Times Biographical Service,* March 1, 1980, 417; Clinton Blake Townsend, "Katharine M[ayor] Townsend," and B. Mayor, "Alfred Goldsborough Mayor," both in *The Alfred G. Mayor and Katharine M. Townsend Memorial Fund* (Woods Hole, Mass.: Woods Hole Oceanographic Institution, 1985), 1–42; interview with Brantz and Ana Mayor by the authors, in Hanover, N.H., July 11–14, 2004.

16. A. G. Mayor to H. H. Mayor, August 11, 1919, in HRHM Papers/SU; A. G. Mayor to Woodward, September 28, 1919, in CIW Archives; *CIW Year Book, 1919,* 18 (1920): 187–89.

17. Alfred Goldsborough Mayor, "Pacific Island Reveries," typescript [1920 or 1921], p. 20, in AGM Papers/SU; C. M. Yonge, *A Year on the Great Barrier Reef: The Story of the Corals and of the Greatest of Their Creations* (London and New York: Putnam, 1930), 99, 101, 170–71; E. W. Gudger, "On the Use of the Diving Helmet in Submarine Biological Work," *American Museum Journal* 18 (1918): 135–38. The diving helmet used at the Tortugas Laboratory was invented by W. S. Dunn, president of the Miller-Dunn Company, Miami, Florida. Gudger's account includes excellent photographs taken underwater in the Tortugas. Superb photographs of Totugas reef fishes in their natural habitat appear in W. H. Longley, "Haunts and Habits of Tropical Fishes," *American Museum Journal* 18 (1918): 78–88.

18. A. G. Mayor, "Structure and Ecology of the Samoan Coral Reefs," 1–25; Alfred Goldsborough Mayor, "Causes Which Produce Stable Conditions in the Depth of the Floors of Pacific Fringing Reef-Flats," *Papers from the Department of Marine Biology of the Carnegie Institution of Washington* 19 (1924): 27–36.

19. *CIW Year Book, 1919,* 185, 191, and *1920,* 19 (1921):185; Alfred Goldsborough Mayor, "The Growth-Rate of Pacific Coral Reefs," *New York Zoological Society Bulletin* 22 (November 1919): 137–39; Alfred Goldsborough Mayor, "The Effect of Diminished Oxygen upon Rate of Nerve Conduction in *Casssiopea,*" *American Journal of Physiology* 51 (April 1, 1920): 543–50; Alfred Goldsborough Mayor, "Samuel Hubbard Scudder, 1837–1911," *National Academy of Sciences Biographical Memoir* 17 (1920): 79–104.

20. *CIW Year Book, 1920,* 19 (1921): 192–93; Woodward to Walcott, May 21, 1920, in Charles D. Walcott Collection, Smithsonian Institution Archives, Washington, D.C.

21. *CIW Year Book, 1920,* 19 (1921): 192–93.

22. A. G. Mayor to Davenport, March 11, 1920, in CBD Papers/APS; *CIW Year Book, 1920,* 186; A. G. Mayor to H. H. Mayor, April 2, 1920, in HRHM Papers/SU.

23. A. G. Mayor to H. H. Mayor, April 28, 1920, in HRHM Papers/SU; A. G. Mayor to Woodward, May 25, 1920, in CIW Archives.

24. A. G. Mayor to H. H. Mayor, June 22, 1920, in HRHM Papers/SU; Alfred Goldsborough Mayor, "Rose Atoll, American Samoa," *Papers from the Department of Marine Biology of the Carnegie Institution of Washington* 19 (1924): 73–79.

25. A. G. Mayor, "Structure and Ecology of the Samoan Coral Reefs," 1–25.

26. Ibid.

27. A. G. Mayor, "Causes Which Produce Stable Conditions," 27–36.

28. A. G. Mayor, "Growth-Rate of Samoan Corals," *Papers from the Department of Marine Biology,* 51–72.

29. Ibid., with photographs of coral specimens.

30. Ibid., 59–60.

31. Alfred Goldsborough Mayor, "The Reefs of Tutuila, Samoa, in Relation to Coral Reef Theories," *Proceedings of the American Philosophical Society* 59 (1920): 224–36; Davis to A. G. Mayor, October 7, 1920, and A. G. Mayor to Rollin T. Chamberlin, [October 7, 1920], in AGM Papers/SU.

32. A. G. Mayor to Davis, October 7, 1920, in AGM Papers/SU.

33. Chamberlin to A. G. Mayor, October 9, 1920, in AGM Papers/SU; *CIW Year Book, 1920,* 185–93.

34. Charles Darwin, *The Structure and Distribution of Coral Reefs,* repr. of 1842 ed., with foreword by Michael T. Ghiselin (Tucson: University of Arizona Press, 1984), 1–79, 88–118; C. M. Yonge, "Darwin and Coral Reefs," in *A Century of Darwin,* ed. S. A. Barnett (Cambridge: Harvard University Press, 1958), 245–66; Dietrich H. H. Kuhlmann, "Darwin's Coral Reef Research—A Review and Tribute," *Marine Ecology* 3, no. 3 (1982): 193–212.

35. Darwin, *Structure and Distribution of Coral Reefs,* 88–118; Yonge, "Royal Society," 438–47.

36. Yonge, "Darwin and Coral Reefs," 263; Yonge, "Royal Society," 443; William A. Herdman, *Founders of Oceanography and Their Work: An Introduction to the Science of the Sea* (London: Edward Arnold & Co., 1923), 114, 205–8. For a thorough treatise on the evolution and distribution of coral reefs, see J. E. N. Veron, *Corals in Space and Time: The Biogeography and Evolution of the Scleractinia* (Ithaca, N.Y.: Cornell University Press, 1995).

37. Norman D. Newell, "Questions of the Coral Reefs," *Natural History* 68 (March 1959): 121–23; Reginald A. Daly, "The Glacial-Control Theory of Coral Reefs," *Proceedings of the American Academy of Arts and Sciences* 51 (1915): 155–251; William Morris Davis, *The Coral Problem,* Special Publication No. 9 (New York: American Geographical Society, 1928), 249, 251.

38. Yonge, "Royal Society," 445; Herdman, *Founders,* 208; Joseph H. Connell, "Population Ecology of Reef-Building Corals," in *Biology and Geology of Coral Reefs,* ed. O. A. Jones and R. Endean (New York and London: Academic Press, 1973), 2:239; Robert Endean, "Destruction and Recovery of Coral Reef Communities," in *Biology and Geology of Coral Reefs,* ed. O. A. Jones and R. Endean (New York and London: Academic Press, 1973), 3:219–21.

39. Woodward to A. G. Mayor, October 30, 1920, in AGM Papers/SU; A. G. Mayor to Cushman, November 19 and 26, and December 14, 1920, in Paleobiology Division, United States National Museum, Washington, D.C.; A. G. Mayor to Davenport, December 19, 1920, in CBD Papers/APS.

40. *CIW Year Book, 1921,* 20 (1922): 187; Reginald A. Daly to A. G. Mayor, January 17, 1921, Conklin to A. G. Mayor, May 25, 1921, and H. H. Mayor to A. G. Mayor, July 3, 1921, in AGM Papers/SU; A. G. Mayor to H. H. Mayor, June 27, 1921, in HRHM Papers/SU; A. G. Mayor to Davenport, January 18, 1921, in CBD Papers/APS.

41. *CIW Year Book, 1921,* 193.

42. Alfred Goldsborough Mayor, "The Tracking Instinct in a Tortugas Ant," *Papers from the Department of Marine Biology of the Carnegie Institution of Washington* 18 (1922): 103–7. Evidence that Mayor had begun to study the ant earlier is found in B. Mayor, "Alfred Goldsborough Mayor," 40–41.

43. A. G. Mayor, "Pacific Island Reveries."

44. A. G. Mayer, "History of Tahiti," 105–7, 403–10, 452–66.

45. A. G. Mayer, "History of Fiji," *Popular Science Monthly* 86 (1915): 521–38.

46. A. G. Mayer, "History of Fiji," *Popular Science Monthly* 87 (1915): 31–49; A. G. Mayer, "History of Fiji," *Popular Science Monthly* 87 (1915): 292–306; A. G. Mayer, "History of Fiji," *Scientific Monthly* 1 (1915): 18–35.

47. Alfred Goldsborough Mayer, "Papua, Where the Stone-Age Lingers," *Scientific Monthly* 1 (1915): 107, 117–18.

48. A. G. Mayer, "Papua," 105–23; A. G. Mayer, "Men of the Mid-Pacific," 5–26.

49. A. G. Mayer, "Men of the Mid-Pacific," 13.

50. A. G. Mayer, "Islands of the Mid-Pacific," 125–48; A. G. Mayer, "Java," 350–54.

51. A. G. Mayor, "Pacific Island Reveries," 2–136.

52. Ibid., 220–25.

53. A. G. Mayor to H. H. Mayor, October 4, 1921, in HRHM Papers/SU; A. G. Mayor to Davenport, October 25, 1921, in CBD Papers/APS.

54. A. G. Mayor to Cushman, December 22, 1921, and March 25, 1922, in Paleobiology Division, United States National Museum, Washington, D.C.; A. G. Mayor to Stewart Paton, March 31, 1922, in AGM Papers/SU.

55. A. G. Mayor to Gilbert, April 4, 1922, in CIW Archives.

56. A. G. Mayor to Dr. J. Alexander Miller, April 17, 1922, and Maria Mayer to A. G. Mayor, May 19, 1922, in AGM Papers/SU; William H. Longley to L. R. Cary, September 20, 1922, in Tortugas Laboratory Papers, American Philosophical Society Library, Philadelphia, Pennsylvania.

57. A. G. Mayor to H. H. Mayor, June 9, 11, and 16, 1922, in HRHM Papers/SU; H. H. Mayor to A. G. Mayor, June 11, 1922, in AGM Papers/SU.

58. Longley to Cary, September 20, 1922, in Tortugas Laboratory Papers, American Philosophical Society Library.

59. Ibid.

60. Ibid.

## Chapter VII—Epilogue

1. *Newark Evening News,* June 28, 1922; *Washington Evening Star,* June 29, 1922, 7; *American Journal of Science,* August 6, 1922; Davenport, *Biographical Memoir,* 8:1–10; T[ownsend], "Alfred Goldsborough Mayor," 138–39; F. A. P[otts], "Dr. A. G. Mayor," *Nature* 110 (August 12, 1922): 224–25.

2. W. A. Setchell to Cushman, October 10, 1922, Longley to Cushman, October 12, 1922, Charles B. Lipman to Cushman, October 20, 1922, E. Newton Harvey to Cushman, October 25, 1922, E. W. Gudger to Cushman, November 9, 1922, Clark to Cushman, November 10, 1922, Ulrich Dahlgren to Cushman, November 16, 1922,

H. H. Mayor to Cushman, November 28, 1922, Paton to Cushman, December 28, 1922, Gilman A. Drew to Cushman, January 8, 1923, Raymond C. Osborn to Cushman, January 8, 1923, H. F. Perkins to Cushman, January 9, 1922 [1923], Edwin Linton to Cushman, January 10, 1923, and Richard M. Field to Cushman, January 14, 1922 [1923]—all in Paleobiology Division, United States National Museum, Washington, D.C.; John W. Mills to Gilbert, July 6, 1923, in CIW Archives; *CIW Year Book, 1923,* 23 (1924): 157; *CIW Minutes of the Meeting of the Executive Committee, Friday, October 27, 1922,* 129, 143. Despite her serious bout with tuberculosis, Harriet Hyatt Mayor lived to the age of ninety-two, dying in 1960, nearly four decades after the passing of her husband (see *New York Times,* December 9, 1960, 31).

3. Joseph J. Betz, "Pioneer Biologist," *Sea Frontiers* 11 (September–October 1965): 287. Both authors of the present work have visited the Tortugas Laboratory site—Dale R. Calder on March 6 and again on March 7, 2004, and Lester D. Stephens on October 13, 2005.

4. Longley to Cary, September 20, 1922, in Tortugas Laboratory Papers, American Philosophical Society Library, Philadelphia, Pennsylvania; "Memorandum for Budget Consideration: Department of Marine Biology," July 11, 1922, and "Memorandum of Conversation with Dr. W. H. Longley," signed by W. M. G[ilbert], November 3, 1922, in CIW Archives; *CIW Year Book, 1923,* 9; John Campbell Merriam, *Published Papers and Addresses of John Campbell Merriam* (Washington, D.C.: Carnegie Institution of Washington, 1938), 4:2516; [William H. Longley], "Memorandum upon Tortugas as the Site of a Tropical Marine Laboratory," typescript [early January 1925], administrative secretary [Gilbert] to J. W. Mills, February 14, 1924, "Schedule of arrivals, for John Mills," March 17, 1924, and W. M. Gilbert, "Memorandum for Investigators, 1924," typescript [undated, but ca. April 1924], in CIW Archives.

5. Longley to Dr. [John Campbell] Merriam, December 16, 1925, and Longley, "Memorandum on Tortugas Laboratory," [early 1926], in CIW Archives.

6. Colin, "Brief History of the Tortugas Marine Laboratory," 145–46.

7. B. Mayor, "Alfred Goldsborough Mayor," 42. In 1940 the Tortugas Laboratory equipment and the *Anton Dohrn* became the property of the Woods Hole Oceanographic Institution (see Colin, "Brief History of the Tortugas Marine Laboratory," 145).

8. Frank M. Truesdale, "Great Invertebrate Zoologists: Alfred Goldsborough Mayor (1868–1922)," *Division of Invertebrate Zoology, American Society of Zoologists Newsletter* (Spring 1993): 9–10.

9. The "Web of Science" is an on-line service of "ISI Web of Knowledge" (The Thomson Corporation).

10. Alfred Goldsborough Mayer, "The Effects of Temperature upon Tropical Marine Animals," *Papers from the Tortugas Laboratory of the Carnegie Institution of Washington* 6 (1914): 1–24, twelve figs.

11. A. G. Mayer, "Rhythmical Pulsation in Scyphomedusae," 1–62, thirty-six figures; Alfred Goldsborough Mayer, "Rhythmical Pulsation in Scyphomedusae—II," *Papers from the Tortugas Laboratory of the Carnegie Institution of Washington* 1 (1908): 113–31, thirteen figs.

12. A. G. Mayor, "Rose Atoll," 73–79; A. G. Mayor, "Structure and Ecology of the Samoan Coral Reefs," 1–25, six plates, two figs.; A. G. Mayor, "Growth-Rate of Samoan Corals," *Papers from the Department of Marine Biology,* 51–72.

13. Citations to works by Alfred Goldsborough Mayer (Mayor), "Web of Science."

# Bibliography

**Manuscript Sources**
(Major collections are indicated by asterisks.)

Academy of Natural Sciences of Philadelphia (Pennsylvania), Ewell Sale Stewart Library,
    Manuscript Collections, Miscellaneous Collections
American Philosophical Society Library, Philadelphia, Pennsylvania
    *Davenport, Charles Benedict, Papers
    Jennings, Herbert Spencer, Papers
    *Tortugas Laboratory Papers
Bancroft Library, University of California at Berkeley
    Ritter, William E., Papers
E. S. Bird Library, Department of Special Collections, Syracuse University, New York
    *Mayer (Mayor), Alfred Goldsborough, Papers
    Mayer, Alfred Marshall, Papers
    *Mayer (Mayor), Harriet Randolph Hyatt, Papers
    Mayer, Maria Snowden, Papers
Brooklyn Museum Archives, Brooklyn, N.Y.
    *Mayer (Mayor), Alfred Goldsborough, Correspondence
Carnegie Institution of Washington Archives, Washington, D.C.
    *Mayer (Mayor), Alfred Goldsborough, Correspondence
Maryland Historical Society Library, Manuscripts Division, Baltimore, Md.
    Miscellaneous collections
Mayor, Brantz and Ana, Private Collection, Hanover, N.H.
    *Mayer (Mayor), Alfred Goldsborough, Papers
    *Mayer (Mayor), Harriet Hyatt, Papers
Museum of Comparative Zoology Library, Harvard University, Cambridge, Mass.
    *Agassiz, Alexander, Letterbook
New York Public Library, Rare Books and Manuscript Division, New York, N.Y.
    *Mayer (Mayor), Alfred Goldsborough, Papers
    *Mayer, Alfred Marshall, Papers

Princeton University Library, Rare Books and Special Collections, Princeton, N.J.
    *Hyatt and Mayer (Mayor) Papers
Smithsonian Institution Archives, Washington, D.C.
    Benjamin, Marcus, Papers
    Rathbun, Richard, Papers
    Vaughan, Thomas Wayland, Papers
    *Walcott, Charles Doolittle, Collection
United States National Museum, Paleobiology Division, Washington, D.C.
    *Cushman, Joseph A., Collection

## Publications of Alfred Goldsborough Mayer (Mayor)

(Listed chronologically by topic)

| ABBREVIATIONS | JOURNAL |
|---|---|
| *Amer. Jour. Phys.* | *American Journal of Physiology* |
| *Amer. Jour. Sci.* | *American Journal of Science* |
| *Amer. Nat.* | *American Naturalist* |
| *Biog. Mem.* | *National Academy of Sciences Biographical Memoirs* |
| *Biol. Bul.* | *Biological Bulletin* |
| *Bul. AMNH* | *Bulletin of the American Museum of Natural History* |
| *Bul. MCZ* | *Bulletin of the Museum of Comparative Zoölogy, at Harvard College* |
| *Bul. U.S. Fish Com.* | *Bulletin of the United States Fish Commission* |
| *Jour. Exper. Zool.* | *Journal of Experimental Zoology* |
| *Mem. MCZ* | *Memoirs of the Museum of Comparative Zoölogy, at Harvard College* |
| *Mem. Nat. Sci.* | *Memoirs of Natural Science, The Museum of the Brooklyn Institute of Arts and Sciences* |
| *Nat. Geog.* | *National Geographic Magazine* |
| *No. Amer. Rev.* | *North American Review* |
| *Papers Dept. Mar. Biol.* | *Papers from the Department of Marine Biology of the Carnegie Institution of Washington* (continuation of *Papers from the Tortugas Laboratory of the Carnegie Institution of Washington;* see below*)* |
| *Papers Tort. Lab.* | *Papers from the Tortugas Laboratory of the Carnegie Institution of Washington (*later, *Papers from the Department of Marine Biology of the Carnegie Institution of Washington)* |
| *Pop. Sci. Mo.* | *Popular Science Monthly* |
| *Proc. APS* | *Proceedings of the American Philosophical Society* |
| *Proc. NAS* | *Proceedings of the National Academy of Sciences* |
| *Proc. 7th Int. Zool. Cong.* | *Proceedings of the Seventh International Zoological Congress* |
| *Proc. Soc. Exper. Biol.* | *Proceedings of the Society for Experimental Biology and Medicine* |
| *Psyche* | *Psyche, A Journal of Entomology* |

| Pub. Car. Inst. Wash. | Publications of the Carnegie Institution of Washington |
|---|---|
| Sci. Bul. | Science Bulletin, The Museum of the Brooklyn Institute of Arts and Sciences |
| Sci. Mo. | Scientific Monthly |
| USNM Bul. | Smithsonian Institution, United States National Museum Bulletin |

*Annelids* (See under *Palolo Worm*)

*Ants*

"The Tracking Instinct in a Tortugas Ant." *Papers Dept. Mar. Biol.* 18 (1922): 101–7.

*Biographical Accounts*

"Alexander Agassiz, 1835–1910." *Pop. Sci. Mo.* 77 (1910): 419–46, portrait and two figures. Reprinted in *Smithsonian Institution Annual Report for 1910.* Washington, D.C.: Smithsonian Institution, 1911, 447–72.

"Alpheus Hyatt, 1838–1902." *Pop. Sci. Mo.* 78 (1911): 129–46, portrait and two figures.

"In Memoriam: George Harold Drew, 1881–1913." *Papers Tort. Lab.* 5 (1914): 1–6.

"William Stimpson, 1832–1872." *Biog. Mem.* 8 (1918): 417–33, portrait.

"Alfred Marshall Mayer, 1836–1897." *Biog. Mem.* 8 (1919): 241–72, portrait. (Co-authored with Robert S. Woodward.)

"Samuel Hubbard Scudder, 1837–1911." *Biog. Mem.* 17 (1920): 79–104, portrait.

*Corals and Coral Reefs* (See also *Temperature, Effects on Cnidarians*)

"An Expedition to the Coral Reefs of Torres Straits." *Pop. Sci. Mo.* 85 (1914): 209–31, illustrations.

"Ecology of the Murray Island Coral Reef." *Proc. NAS* 1 (1915): 211–14.

"Sub-marine Solution of Limestone in Relation to the Murray-Agassiz Theory of Coral Atolls." *Proc. NAS* 2 (1916): 28–30.

"Coral Reefs of Tutuila, with Reference to the Murray Agassiz Solution Theory." *Proc. NAS* 3 (1917): 522–26.

"Ecology of the Murray Island Coral Reef." *Papers Dept. Mar. Biol.* 9 (1918): 3–48.

"The Growth-Rate of Samoan Coral Reefs." *Proc. NAS* 4 (1918): 390–93.

"The Growth-Rate of Pacific Coral Reefs." *New York Zoological Society Bulletin* 22 (November 1919): 137–39.

"The Reefs of Tutuila, Samoa, in Relation to Coral Reef Theories." *Proc. APS* 59 (1920): 224–36, three figures.

"Rose Atoll, American Samoa." *Papers Dept. Mar. Biol.* 19 (1924): 73–79.

"Structure and Ecology of the Samoan Coral Reefs." *Papers Dept. Mar. Biol.* 19 (1924): 1–25, six plates, two figures.

"Inability of Stream-Water to Dissolve Submarine Limestones." *Papers Dept. Mar. Biol.* 19 (1924): 37–49, one figure.

"Growth-Rate of Samoan Corals." *Papers Dept. Mar. Biol.* 19 (1924): 51–72, twenty-six plates.

"Causes Which Produce Stable Conditions in the Depth of the Floors of Pacific Fringing Reef-Flats." *Papers Dept. Mar. Biol.* 19 (1924): 27–36, two plates, one figure.

*Ctenophores (or Comb Jellies)*

*Ctenophores of the Atlantic Coast of North America. Pub. Car. Inst. Wash.*, no. 163. Washington, D.C.: Carnegie Institution of Washington, 1912, fifty-eight pages, seventeen plates, twelve figures.

*Hydromedusae (*See under *Medusae)*

*Lepidoptera*

"The Development of the Wing Scales and Their Pigment in Butterflies and Moths." *Bul. MCZ* 29 (1896): 209–36, seven plates.

"On the Color and Color-Patterns of Moths and Butterflies." *Bul. MCZ* 30 (1897): 169–256, ten plates.

"A New Hypothesis of Seasonal Dimorphism in Lepidoptera." *Psyche* 8 (1897): 47–50, 59–62.

"On the Development of Color in Moths and Butterflies." *Biological Lectures from the Marine Biological Institute of Woods Holl, 10th lecture, 1889* (Woods Hole, Mass.: Marine Biological Laboratory, 1900), 157–64.

"On the Mating Instinct in Moths." *Annals and Magazine of Natural History,* 7th ser., 5 (1900): 183–90. Also in *Psyche* 9 (1900): 15–20.

"Effects of Natural Selection and Race-Tendency upon the Color-Patterns of Lepidoptera." *Sci. Bul.* 1 (1902): 31–86, two plates.

"Some Reactions of Caterpillars and Moths." *Jour. Exper. Zool.* 3 (1906): 415–33. (Coauthored with Caroline G. Soule.)

*Marine Life*

*Sea-shore Life: The Invertebrates of the New York Coast and the Adjacent Coast Region.* New York: New York Zoological Society, 1905, 1–181. Reprinted in 1906, 1911, and 1916.

"On the Non-existence of Nervous Shell-Shock in Fishes and Marine Invertebrates." *Proc. NAS* 3 (1917): 597–98.

*Medusae*

"An Account of Some Medusae Obtained in the Bahamas." *Bul. MCZ* 25 (1894): 235–41, three plates.

"On Some Medusae from Australia." *Bul. MCZ* 32 (1898): 15–19, three plates. (Coauthored with Alexander Agassiz.)

"On *Dactylometra.*" *Bul. MCZ* 32 (1898): 1–11, thirteen plates. (Coauthored with Alexander Agassiz.)

"Acalephs from the Fiji Islands." *Bul. MCZ* 32 (1899): 157–89, seventeen plates, 146 figures. (Coauthored with Alexander Agassiz.)

"Descriptions of New and Little-known Medusae from the Western Atlantic." *Bul. MCZ* 37 (1900): 1–9, six plates.

"Some Medusae from the Tortugas, Florida." *Bul. MCZ* 37 (1900): 13–82, forty-four plates.

"The Variations of a Newly-Arisen Species of Medusa." *Sci. Bul.* 1 (1901): 1–27, two plates.

"Medusae [of the Tropical Pacific]." *Mem. MCZ* 26 (1902): 139–75, thirteen plates, one map. (Coauthored with Alexander Agassiz.)

"Medusae of the Bahamas." *Mem. Nat. Sci.* 1 (1904): 1–33, seven plates.

"Medusae of the Hawaiian Islands Collected by the Steamer *Albatross* in 1902." *Bul. U.S. Fish Com. for 1903,* 23, pt. 3 (1904): 1131–43, three plates.

"Rhythmical Pulsation in Scyphomedusae." *Pub. Car. Inst. Wash.,* no. 47. Washington, D.C.: Carnegie Institution of Washington, 1906, 1–62, thirty-six figures.

"Rhythmical Pulsation in Scyphomedusae—II." *Papers Tort. Lab.* 1 (1906): 113–31, thirteen figures.

"The Relation between Ciliary and Muscular Movements." *Proc. Soc. Exper. Biol.* 7 (1909): 19–20.

*Medusae of the World.* 3 vols. Pub. 109. Washington, D.C.: Carnegie Institution of Washington, 1910, 735 pages, seventy-six plates, 428 figures. Reprinted 1977.

"The Converse Relation between Ciliary and Neuro-muscular Movements." *Papers Tort. Lab.* 3 (1911): 1–25, eight figures.

"The Cause of Rhythmical Pulsation in Scyphomedusae." *Proc. 7th Int. Zool. Cong.* [August 19–24, 1907], 1912, 278–81 (updated in 1908 before publication in 1912).

"The Relation between the Degree of Concentration of the Electrolytes of Sea-Water and the Rate of Nerve-Conduction in *Cassiopea*." *Papers Tort. Lab.* 6 (1914): 25–54.

"The Law Governing the Loss of Weight in Starving *Cassiopea*." *Papers Tort. Lab.* 6 (1914): 55–82, one plate, twenty-one figures.

"The Nature of Nerve Conduction in *Cassiopea*." *Proc. NAS* 1 (1915): 270–74.

"Medusae of the Philippines and of Torres Straits." *Pub. Car. Inst. Wash.,* no. 212. Washington, D.C.: Carnegie Institution of Washington, 1915, 157–203, three plates, seven figures.

"A Theory of Nerve-Conduction." *Proc. NAS* 2 (1916): 37–41, two figures.

"Nerve Conduction, and Other Reactions in *Cassiopea*." *Amer. Jour. Phys.* 39 (1916): 375–93, two figures.

"Further Studies of Nerve Conduction in *Cassiopea*." *Proc. NAS* 2 (1916): 721–26, two figures.

"Report upon the Scyphomedusae Collected by the United States Bureau of Fisheries Steamer 'Albatross' in the Philippine Islands and Malay Archipelago." *USNM Bul.* 100 (1917): 173–233. Revised version of "Medusae of the Philippines and of Torres Straits" (see above).

"Further Studies of Nerve Conduction in *Cassiopea*." *Amer. Jour. Sci.* 42 (1917): 469–75, two figures.

"Nerve Conduction in *Cassiopea xamachana*." *Papers Dept. Mar. Biol.* 11 (1917): 1–20, fifteen figures.

"Formula for Rate of Nerve Conduction in Sea Water." *Amer. Jour. Phys.* 44 (1917): 591–95.

"Nerve-Conduction in Diluted and in Concentrated Sea-water." *Papers Dept. Mar. Biol.* 12 (1918): 179–83, one figure.

"The Effect of Diminished Oxygen upon Rate of Nerve Conduction in *Cassiopea*."
*Amer. Jour. Phys.* 51 (1920): 543–50.

*Miscellaneous*
"The Radiation and Absorption of Heat by Leaves." *Amer. Jour. Sci.* 45 (1893): 340–46,
one figure.
"On an Improved Heliostat Invented by Alfred M. Mayer."*Amer. Jour. Sci.,* 4th ser., 4
(1897): 306–8, two figures.
"Our Neglected Southern Coast." *Nat. Geog.* 19 (1908): 859–71.
["The Mission of Scientists."] Address of the Retiring Vice President of Section F of the
American Association for the Advancement of Science. *Science* 61 (1915): 81–82.

*Mollusks*
"Some Species of *Partula* from Tahiti: A Study in Variation." *Mem. MCZ* 26 (1902):
117–35, one plate.

*Museums*
"Educational Efficiency of Our Museums." *No. Amer. Rev.* 177 (1903): 564–69.
"The Status of Public Museums in the United States." *Science,* n.s., 17 (1903): 843–51.

*Navigation*
*Navigation, Illustrated by Diagrams.* Philadelphia and London: J. B. Lippincott, 1918,
1– 207, ninety-seven figures.

*Palolo Worm*
"An Atlantic 'Palolo,' *Staurocephalus gregaricus.*" *Bul. MCZ* 36 (1900): 1–14, three
plates.
"The Atlantic Palolo." *Sci. Bul.* 1 (1902): 93–103, one plate.
"The Annual Breeding-Swarm of the Atlantic Palolo." *Pub. Car. Inst. Wash.,* no. 102,
Washington, D.C.: Carnegie Institution of Washington, 1908, 105–12, one plate.
"The Annual Swarming of the Atlantic Palolo." *Proc. 7th Int. Zool. Cong.* [August 19–
24, 1907] (1912): 147–151 (updated through 1908 before publication in 1912).

*Peoples of Oceania*
"A History of Fiji." *Pop. Sci. Mo.* 86 (1915): 521–38; 87 (1915): 31–49, 292–306,
photographs; *Sci. Mo.* 1 (1915):18–35, photographs.
"A History of Tahiti." *Pop. Sci. Mo.* 86 (1915): 105–27, 403–10, 452–66, photographs.
"Papua, Where the Stone-Age Lingers." *Sci. Mo.* 1 (1915): 105–23, photographs.
"The Islands of the Mid-Pacific." *Sci. Mo.* 2 (1916): 125–48, photographs.
"Java, the Exploited Island." *Sci. Mo.* 2 (1916): 350–54.
"The Men of the Mid-Pacific." *Sci. Mo.* 2 (1916): 5–26, photographs.

*Preparation of Specimens*
"On the Use of Carbon Dioxide in Killing Marine Animals." *Biol. Bul.* 16 (1908): 18.
"On the Use of Magnesium in Stupefying Marine Animals." *Biol. Bul.* 17 (1909): 341–
42.

*Reptiles*
"Habits of the Box Tortoise." *Pop. Sci. Mo.* 38 (1890): 60–65, three figures.
"Habits of the Garter Snake." *Pop. Sci. Mo.* 42 (1893): 485–88, four figures.

*Scyphozoa (*See under *Medusae)*

*Sea Water*

"The Gulf Stream off Our Coast." (Cited in Charles B. Davenport's list of Mayor's works as *Bul. AMNH* 16 [1916]: 501–6, but does not appear there. Not located by present authors.)

"Observations upon the Alkalinity of the Surface Water of the Tropical Pacific." *Proc. NAS* 3 (1917): 548–52.

"Detecting Ocean Currents by Observing Their Hydrogen-ion Concentration." *Proc. APS* 58 (1919): 150–60, one figure.

"Hydrogen-ion Concentration and Electrical Conductivity of the Surface Water of the Atlantic and Pacific." *Papers Dept. Mar. Biol.* 18 (1922): 61–85, three charts.

*Temperature, Effects on Cnidarians*

"The Effects of Temperature upon Tropical Marine Animals." *Papers Tort. Lab.* 6 (1914): 1–24, twelve figures.

"Is Death from High Temperature Due to the Accumulation of Acid in the Tissues?" *Amer. Jour. Phys.* 44 (1917): 581–85. Also in *Proc. NAS* 3 (1917): 626–27.

"Toxic Effects Due to High Temperature." *Papers Dept. Mar. Biol.* 12 (1918): 173–78.

*Tortugas Laboratory*

"The Tortugas, Florida, as a Station for Research in Biology." *Science,* n.s., 17 (1903): 190–92.

"A Tropical Marine Laboratory for Research?" *Science,* n.s., 17 (1903): 655–60.

"A Laboratory for the Study of Marine Zoology in the Tropical Atlantic." *Pop. Sci. Mo.* 64 (1903): 41–47, six figures.

"The Bahamas vs. Tortugas as a Station for Research in Marine Zoology." *Science,* n.s., 18 (1903): 369–71.

"A Plan for Increasing the Efficiency of Marine Expeditions." *Science,* n.s., 27 (1908): 669–71.

"Marine Laboratories, and Our Atlantic Coast." *Amer. Nat.* 42 (1908): 533–36.

"The Research Work of the Tortugas Laboratory." *Pop. Sci. Mo.* 76 (1910): 396–411, twelve figures.

"The Tortugas Laboratory of the Carnegie Institution of Washington." *Internationale Revue der gesamten Hydrobiologie und Hydrographie* 5 (1913): 505–10.

Annual Reports to the Carnegie Institution of Washington, *Carnegie Institution of Washington Yearbook,* no. 3, *1904* (1905): 50–54, two plates; no. 4, *1905* (1906): 108–24, two plates; no. 5, *1906* (1907): 106–18, one plate; no. 6, *1907* (1908): 106–23, one plate; no. 7, *1908* (1909): 118–38, one plate; no. 8, *1909* (1910): 125–53, one plate; no. 9, *1910* (1911): 117–48, one plate; no. 10, *1911* (1912): 120–56, four plates; no.11, *1912* (1913): 118–64, two plates; no. 12, *1913* (1914): 163–84; no.13, *1914* (1915): 169–233; no.14, *1915* (1916): 183–241; no. 15, *1916* (1917): 171–215; no.16, *1917* (1918): 161–89; no. 17, *1918* (1919): 149–72; no.18, *1919* (1920): 185–210; no.19, *1920* (1921): 185–200; no. 20, *1921* (1922): 191–205.

*Universities*

"Material versus Intellectual Development of Our Universities." *Science*, n.s., 20 (1904): 44–47.

"Should Our Colleges Establish Summer Schools?" *Science*, n.s., 23 (1906): 703–4.

"Autonomy for the University?" *Science*, n.s., 30 (1909): 673–75.

"Our Universities and Research." *Science*, n.s., 32 (1910): 257–60.

**Other Sources**

Agassiz, Alexander. Letter to Dr. Billings. In *Carnegie Institution of Washington Yearbook, 1902*. Washington, D.C.: Carnegie Institution of Washington, 1903.

"Alfred Marshall Mayer." In *Appleton's Cyclopaedia of American Biography*, 6 vols. (New York: D. Appleton and Co., 1889–1900), 4:274.

Allen, Garland E. "Davenport, Charles Benedict." In *American National Biography*, 24 vols. (New York: Oxford University Press, 1999), 6:126–28.

———. "Heredity, Development, and Evolution." In *Centennial History of the Carnegie Institution of Washington*, vol. 5: *The Department of Embrology*, ed. Jane Maienschein, Marie Glitz, and Garland E. Allen, 145–72. Cambridge: Cambridge University Press, 2004.

Anonymous. "A Monograph of the Jellyfishes." *Nature* 85 (1910): 285–87.

Appel, Toby A. "Organizing Biology: The American Society of Naturalists and Its 'Affiliated Societies,' 1883–1923." In *The American Development of Biology*, ed. Ronald Rainger, Keith R. Benson, and Jane Maienschein, 87–120. Philadelphia: University of Pennsylvania Press, 1988.

Arai, Mary N. *A Functional Biology of Scyphozoa*. London: Chapman and Hall, 1997.

Arai, Mary N., and Anita Brinckmann-Voss. "Hydromedusae of British Columbia and Puget Sound." *Canadian Bulletin of Fisheries and Aquatic Sciences* 204 (1980): 1–192.

Bakus, Gerald J. "The Biology and Ecology of Tropical Holothurians." In *Biology and Geology of Coral Reefs*, ed. O. A. Jones and R. Endean, 2:325–67. New York and London: Academic Press, 1973.

Benson, Keith R. "From Museum Research to Laboratory Research: The Transformation of Natural History into Academic Biology." In *The American Development of Biology*, ed. Ronald Rainger, Keith R. Benson, and Jane Maienschein, 49–83. Philadelphia: University of Pennsylvania Press, 1988.

———. "Laboratories on the New England Shore: 'The Somewhat Different Direction' of American Marine Biology." *New England Quarterly* 61 (March 1988): 55–78.

———. "Review Paper: The Naples Stazione Zoologica and Its Impact on the Emergence of American Marine Biology." *Journal of the History of Biology* 21 (Summer 1988): 331–41.

Betz, Joseph J. "Pioneer Biologist." *Sea Frontiers* 11 (September–October 1965): 286–95.

Bouillon, J., and F. Boero. "Synopsis of the Families and Genera of the Hydromedusae of the World, with a List of the Worldwide Species." *Thalassia Salentina* 24 (2000): 47–296.

Brooklyn Institute of Arts and Sciences. *Prospectus for 1900–1901.* Brooklyn: The Institute, 1900.

———. *Thirteenth Year Book of the Brooklyn Institute of Arts and Sciences.* Brooklyn: The Institute, 1901.

Buckley, Kerry W. *Mechanical Man: John Broadus Watson and the Beginnings of Behaviorism.* New York: The Guilford Press, 1989.

Calder, Dale R. *Shallow-Water Hydroids of Bermuda: The Thecatae, Exclusive of Plumularioidea.* Royal Ontario Museum Life Sciences Contributions, 154. Toronto: Royal Ontario Museum, 1991.

Carnegie Institution of Washington. *Minutes of the Meeting of the Executive Committee, Thursday, January 18, 1917.* Washington, D.C.: Carnegie Institution of Washington, 1917.

———. *Minutes of the Meeting of the Executive Committee, Friday, October 27, 1922.* Washington, D.C.: Carnegie Institution of Washington, 1922.

Clark, Hubert Lyman. *Carnegie Scientists in the Antipodes.* Boston, 1914, 1–6.

———. Letter to Editor. *Science,* n.s., 17 (June 19, 1903): 979–80.

Clark, Leonard B., and Walter N. Hess. "Swarming of the Atlantic Palolo Worm, *Leodice fucata* (Ehlers)." *Papers from the Marine Biological Laboratory of the Carnegie Institution of Washington* 33 (1940): 23–33.

Colin, P. L. "A Brief History of the Tortugas Marine Laboratory." In *Oceanography: The Past,* ed. M. Sears and D. Merriam, 138–47. New York: Springer-Verlag, 1980.

Connell, Joseph H. "Population Ecology of Reef-Building Corals." In *Biology and Geology of Coral Reefs,* ed. O. A. Jones and R. Endean, 2:205–45. New York and London: Academic Press, 1973.

Cornelius, P. F. S. "Donat Vladimirovitch Naumov (1921–1984)." In *Modern Trends in the Systematics, Ecology, and Evolution of Hydroids and Hydromedusae,* ed. J. Bouillon, F. Boero, F. Cicogna, and P. F. S. Cornelius, 5–7. Oxford: Oxford University Press, 1987.

Cravens, Hamilton. *The Triumph of Evolution: American Scientists and the Heredity-Environment Controversy, 1900–1941.* Philadelphia: University of Pennsylvania Press, 1978.

Dahl, A. L., and A. E. Lamberts. "Environmental Impact on a Samoan Coral Reef: A Resurvey of Mayor's 1917 Transect." *Pacific Science* 31 (July 1977): 309–19.

Daly, Reginald A. "The Glacial-Control Theory of Coral Reefs." *Proceedings of the American Academy of Arts and Sciences* 51 (1915): 155–251.

Darwin, Charles. *The Structure and Distribution of Coral Reefs.* Reprint of 1842 ed., with foreword by Michael T. Ghiselin. Tucson: University of Arizona Press, 1984.

Davenport, Charles B. *Biographical Memoir of Alfred Goldsborough Mayor, 1868–1922.* Washington, D.C.: National Academy of Sciences, 1926.

Davis, John H., Jr. "The Ecology of the Vegetation and Topography of the Sand Keys of Florida." *Papers from Tortugas Laboratory* 33 (November 1942): 121–91.

Davis, William Morris. *The Coral Problem.* Special Publication No. 9. New York: American Geographical Society, 1928.

Dexter, Ralph W. "The Annisquam Sea-side Laboratory of Alpheus Hyatt, Predecessor of the Marine Biological Laboratory at Woods Hole, 1880–1886." In *Oceanography: The Past,* ed. M. Sears and D. Merriam, 94–100. New York: Springer-Verlag, 1980.

Eastman, Rebecca Hooper. *The Story of the Brooklyn Institute of Arts and Sciences, 1824–1924.* N.p., [1925].

Ebert, James D. "Carnegie Institution of Washington and Marine Biology: Naples, Woods Hole, and Tortugas." *Biological Bulletin* 168 (June 1985): 172–82.

Endean, Robert. "Destruction and Recovery of Coral Reef Communities." In *Biology and Geology of Coral Reefs,* ed. O. A. Jones and R. Endean, 3:215–54. New York: Academic Press, 1976.

Fisher, M. M., and John J. Rice. *History of Westminster College, 1851–1903.* Columbia, Mo.: Press of E. W. Stephens, 1908.

Graham, W. M., D. L. Martin, D. L. Felder, V. L. Asper, and H. M. Perry. "Ecological and Economic Implications of a Tropical Jellyfish Invader in the Gulf of Mexico." *Biological Invasions* 5 (2003): 53–69.

Gudger, E. W. "On the Use of the Diving Helmet in Submarine Biological Work." *American Museum Journal* 18 (February 1918): 135–38.

Haeckel, Ernst. *Das System der Medusen: Erster Theil einer Monographie der Medusen.* Jena: Verlag von Gustav Fischer, 1879.

———. *System der Acraspeden: Zweite Hälfte des Systems der Medusen.* Jena: Verlag von Gustav Fischer, 1880.

Halstead, B. W. *Poisonous and Venomous Marine Animals of the World.* 2nd rev. ed. Princeton, N.J.: Darwin Press, 1988.

Harbison, G. Richard, and Laurence P. Madin. "An Appreciation of Alfred G. Mayor." In *The Alfred G. Mayor and Katharine M. Townsend Memorial Fund,* 45–49. Woods Hole, Mass.: Woods Hole Oceanographic Institution, 1985.

"Harriet Hyatt Mayer." *Brookgreen Bulletin* (Summer 1973): n.p.

Hefelbower, Samuel Gring. *The History of Gettysburg College, 1833–1932.* Gettysburg, Pa.: Gettysburg College, 1932.

Herdman, William A. *Founders of Oceanography and Their Work: An Introduction to the Science of the Sea.* London: Edward Arnold & Co., 1923.

Hiltzik, Lee Richard. "The Brooklyn Institute of Arts and Sciences' Biological Laboratory, 1890–1924: A History." Ph.D. diss., State University of New York at Stonybrook, 1993.

"Hooper, Franklin William." In *National Cyclopedia of American Biography,* 63 vols. to date (New York: J. T. White, 1891– ), 13:46–47.

Hurley, Neil E. *Lighthouses of the Dry Tortugas, An Illustrated History.* Aiea, Hawaii: Historic Lighthouse Publishers, 1994.

Kofoid, Charles A. "The Biological Stations of Europe." *United States Bureau of Education Bulletin, 1910,* no. 4 (Washington, D.C.: Government Printing Office, 1910), i–xiii, 1–360.

Kohlstedt, Sally Gregory. "Museums on Campus: A Tradition of Inquiry and Learning." In *The American Development of Biology,* ed. Ronald Rainger, Keith R. Benson, and Jane Maienschein, 15–47. Philadelphia: University of Pennsylvania Press, 1988.

Kramp, Paul L. "The Hydromedusae of the Atlantic Ocean and Adjacent Waters." *Dana-Report* 46 (1959): 1–283.

———. "The Hydromedusae of the Pacific and Indian Oceans." *Dana-Report* 72 (1968): 1–200.

———. "Synopsis of the Medusae of the World." *Journal of the Marine Biological Association of the United Kingdom* 40 (1961): 1–469.

Krukenberg, C. F. W. "Über den Wassergehalt der Medusen." *Zoologischer Anzeiger* 3 (1880): 306.

Kuhlmann, Dietrich H. H. "Darwin's Coral Reef Research—A Review and Tribute." *Marine Ecology* 3, no. 3 (1982): 193–212.

Landrum, L. Wayne. *Fort Jefferson and the Dry Tortugas National Park.* Big Pine Key, Fla.: the author, 2003.

Lillie, Frank R. *The Woods Hole Marine Biological Laboratory.* Chicago: University of Chicago Press, 1944.

Longley, W. H. "Haunts and Habits of Tropical Fishes." *American Museum Jounal* 18 (February 1918): 78–88.

Lurie, Edward. *Louis Agassiz: A Life in Science.* Chicago: University of Chicago Press, 1960.

Maienschein, Jane. "Agassiz, Hyatt, Whitman, and the Birth of the Marine Biological Laboratory." *Biological Bulletin* 6 (1985): 26–34.

———. "Introduction." In *Centennial History of the Carnegie Institution of Washington,* vol. 5: *The Department of Embrology,* ed. Jane Maienschein, Marie Glitz, and Garland E. Allen, 1–20. Cambridge: Cambridge University Press, 2004.

———. *One-Hundred Years Exploring Life, 1888–1988: The Marine Biological Laboratory at Woods Hole.* Boston: Jones and Bartlett Publishers, 1989.

———. *Transforming Traditions in American Biology, 1880–1915.* Baltimore: Johns Hopkins University Press, 1991.

Marques, A. C., and A. G. Collins. "Cladistic Analysis of Medusozoa and Cnidarian Evolution." *Invertebrate Biology* 123 (2004): 23–42.

Mayer, Alfred M. "Eulogy on Joseph Henry." *Proceedings of the American Association for the Advancement of Science* 29 (1881): 69.

Maÿer, Harriet Hyatt. *The Maÿer Family.* N.p., [1911].

Mayor, Brantz. "Alfred Goldsborough Mayor." In *The Alfred G. Mayor and Katharine M. Townsend Memorial Fund,* 29–42. Woods Hole, Mass.: Woods Hole Oceanographic Institution, 1985.

Merriam, John Campbell. *Published Papers and Addresses of John Campbell Merriam.* Vol. 4. Washington, D.C.: Carnegie Institution of Washington, 1938.

Miller, Howard S. *Dollars for Research: Science and Its Patrons in Nineteenth-Century America.* Seattle: University of Washington Press, 1970.

Mills, Claudia E. "Jellyfish Blooms: Are Populations Increasing Globally in Response to Changing Ocean Conditions?" *Hydrobiologia* 451 (2001): 55–68.

Mills, Eric L. *Biological Oceanography: An Early History, 1870–1960.* Ithaca, N.Y.: Cornell University Press, 1989.

Naumov, D. V. "Gidroidy I Gidromeduzy Morskikh, Solonovatovodnykh I Presnovodnykh Basseinov SSSR." *Akademiya Nauk SSSR, Opredeliteli po Faune SSSR* 70

(1960): 1–626. (Translated as *Hydroids and Hydromedusae of the USSR.* Jerusalem: Israel Program for Scientific Translations, 1969, Catalog Number 5108, 1–660.)

———. "Stsifoidnye Meduzy Morei SSSR." *Akademiya Nauk SSSR, Opredeliteli po Faune SSSR* 75 (1961): 1–98.

*New York Times,* June 16, 1894; July 14, 1897; March 5, 1905; February 18, 1906; October 9, 1910; and December 9, 1960.

*New York Times Biographical Service,* March 1, 1980, 417–18.

*New York Tribune,* July 14, 1897.

*Newark Evening News,* June 28, 1922.

Newell, Norman D. "Questions of the Coral Reefs." *Natural History* 68 (March 1959): 121–23.

Numbers, Ronald L. *Darwinism Comes to America.* Cambridge: Harvard University Press, 1998.

Nutting, Charles C. "American Hydroids, Part II: Sertularidae." *Smithsonian Institution, United States National Museum Special Bulletin* no. 4 (1901): 1–325.

———. Review of A. G. Mayer's *Medusae of the World,* vols. 1 and 2. *Science,* n.s., 32 (1910): 596–97.

[Obituary of Alfred G. Mayor.] *American Journal of Science,* 5th ser., 4 (August 1922): 173–74.

Omori, M., and E. Nakano. "Jellyfish Fisheries in Southeast Asia." *Hydrobiologia* 451 (2001): 19–26.

Passano, L. M. "Scyphozoa and Cubozoa." In *Electrical Conduction and Behaviour in "Simple" Invertebrates,* ed. G. A. B. Shelton, 149–202. Oxford: Clarendon Press, 1982.

P[otts], F. A. "Dr. A. G. Mayor." *Nature* 110 (August 12, 1922): 224–25.

Purcell, Jennifer E., and Mary N. Arai. "Interactions of Pelagic Cnidarians and Ctenophores with Fish: A Review." *Hydrobiologia* 451 (2001): 27–44.

Purcell, J. E., W. M. Graham, and H. J. Dumont. "Jellyfish Blooms: Ecological and Societal Importance." *Hydrobiologia* 451 (2001): 333pp.

Raitt, Helen, and Beatrice Moulton. *Scripps Institution of Oceanography: First Fifty Years.* N.p.: The Ward Ritchie Press, 1967.

Rolfs, P. H. Letter to Editor. *Science,* n.s., 17 (June 26, 1903): 1008–9.

Romanes, George John. "The Croonian Lecture: Preliminary Observations on the Locomotor System of Medusae." *Philosophical Transactions of the Royal Society of London* 166 (1876): 269–313.

———. "Further Observations on the Locomotor System of Medusae." *Philosophical Transactions of the Royal Society of London* 167 (1877): 659–752.

———. "Concluding Observations on the Locomotor System of Medusae." *Philosophical Transactions of the Royal Society of London* 171 (1880): 161–202.

Russell, F. S. *The Medusae of the British Isles: Anthomedusae, Leptomedusae, Limnomedusae, Trachymedusae and Narcomedusae.* Cambridge: Cambridge University Press, 1953.

———. *The Medusae of the British Isles, II: Pelagic Scyphozoa with a Supplement to the First Volume on Hydromedusae.* Cambridge: Cambridge University Press, 1970.

Saville-Kent, William. *The Great Barrier Reef of Australia: Its Products and Potentialities.* London: W. H. Allen and Company, [1893].

Schmidt, T. W., and L. Pikula. *Scientific Studies on Dry Tortugas National Park: An Annotated Bibliography.* National Park Service, Current References 97-1, 1–108. Washington, D.C.: U.S. Department of Commerce, National Oceanic and Atmospheric Administration, National Oceanographic Data Center, and Department of the Interior, 1997.

Shiras, George III. "One Season's Game-Bag with the Camera." *National Geographic Magazine* 19 (June 1908): 388–402.

Stephens, Lester D., and Dale R. Calder. "John McCrady of South Carolina: Pioneer Student of North American Hydrozoa." *Archives of Natural History* 19 (1992): 39–54.

T. B. Review of *Sea-shore Life. American Naturalist* 40 (May 1906): 378.

T[ownsend], C[harles] H. "Alfred Goldsborough Mayor." *New York Zoological Society Bulletin* 25 (November 1922): 138.

Townsend, Clinton Blake. "Katharine M[ayor] Townsend." In *The Alfred G. Mayor and Katharine M. Townsend Memorial Fund,* 5–26. Woods Hole, Mass.: Woods Hole Oceanographic Institution, 1985.

Trefil, James, and Margaret Hindle Hazen. *Good Seeing: A Century of Science at the Carnegie Institution of Washington, 1902–2002.* Washington, D.C.: Joseph Henry Press, 2002.

Truesdale, Frank M. "Great Invertebrate Zoologists: Alfred Goldsborough Mayor (1868–1922)." *Division of Invertebrate Zoology, American Society of Zoologists Newsletter* (Spring 1993): 5–10.

Vasile, Ronald S. "Stimpson, William." In *American National Biography Online.*

Vaughan, T. Wayland. "Corals and the Formation of Coral Reefs." In *Annual Report of the Smithsonian Institution, 1917,* 189–238. Washington, D.C.: The Smithsonian Institution, 1919.

———. "Geology of the Florida Keys and the Marine Deposits and Recent Corals of Southern Florida," *CIW Year Book, 1908,* 7 (1909): 131–6.

Veron, J. E. N. *Corals in Space and Time: The Biogeography and Evolution of the Scleractinia.* Ithaca, N.Y.: Cornell University Press, 1995.

*Washington Evening Star,* June 29, 1922.

Williamson, J. A., P. F. Fenner, J. W. Burnett, and J. F. Rifkin, eds. *Venomous and Poisonous Marine Animals: A Medical and Biological Handbook.* Sydney: University of New South Wales Press, 1996.

Winsor, Mary P. *Reading the Shape of Nature: Comparative Zoology at the Agassiz Museum.* Chicago: University of Chicago Press, 1991.

Wood Jones, Frederic. "On the Growth-forms and Supposed Species in Corals." *Proceedings of the Zoological Society of London* 31 (June 18, 1907): 518–56.

Yonge, C. M. "Darwin and Coral Reefs." In *A Century of Darwin,* ed. S. A. Barnett, 245–66. Cambridge: Harvard University Press, 1958.

———. "Development of Marine Biological Laboratories." *Science Progress* 64 (January 1956): 1–15.

———. "The Royal Society and the Study of Coral Reefs." In *Oceanography: The Past*, ed. M. Sears and D. Merriam, 438–47. New York: Springer-Verlag, 1980.

———. *A Year on the Great Barrier Reef: The Story of the Corals and of the Greatest of Their Creations*. London and New York: Putnam, 1930.

Index

# About the Authors

Professor of history emeritus at the University of Georgia, LESTER D. STEPHENS is the author of numerous works on the lives and work of American naturalists, including *Joseph LeConte, Gentle Prophet of Evolution* and *Science, Race, and Religion in the American South: John Bachman and the Charleston Circle of Naturalists, 1815–1895*.

Marine biologist DALE R. CALDER is curator emeritus in the Department of Natural History at the Royal Ontario Museum; associate professor of zoology at the University of Toronto; and a research associate of the Bermuda Aquarium, Museum, and Zoo. He has served for the past six years on the International Commission on Zoological Nomenclature.